535840

D6
37.50

Butterworths Food La

To my wife

Butterworths Food Law

AA Painter
Editor of *Butterworths Law of Food and Drugs* and *O'Keefe's Law of Weights and Measures*

Butterworths
London, Dublin & Edinburgh
1992

United Kingdom	Butterworth & Co (Publishers) Ltd, 88 Kingsway, LONDON WC2B 6AB and 4 Hill Street, EDINBURGH EH2 3JZ
Australia	Butterworths Pty Ltd, SYDNEY, MELBOURNE, BRISBANE, ADELAIDE, PERTH, CANBERRA and HOBART
Belgium	Butterworth & Co (Publishers) Ltd, BRUSSELS
Canada	Butterworths Canada Ltd, TORONTO and VANCOUVER
Ireland	Butterworth (Ireland) Ltd, DUBLIN
Malaysia	Malayan Law Journal Sdn Bhd, KUALA LUMPUR
New Zealand	Butterworths of New Zealand Ltd, WELLINGTON and AUCKLAND
Puerto Rico	Equity de Puerto Rico, Inc, HATO REY
Singapore	Malayan Law Journal Pte Ltd, SINGAPORE
USA	Butterworth Legal Publishers, AUSTIN, Texas; BOSTON, Massachusetts; CLEARWATER, Florida (D & S Publishers); ORFORD, New Hampshire (Equity Publishing); ST PAUL, Minnesota; and SEATTLE, Washington

A CIP Catalogue record for this book is available from the British Library.

ISBN 0 406 00642 3

Disc conversion by Columns Design and Production Services Ltd, Reading.
Printed by Billings Book Plan, Worcester.

Preface

As a consultant on food law, and editor of *Butterworths Law of Food and Drugs* and *O'Keefe's Law of Weights and Measures*, it has been apparent to me for some time that there is a need for a concise book on food law drawing attention to new developments, discussing matters of particular difficulty and identifying all the legislation applying to food. Quite independently, my colleagues in Butterworth Law Publishers Ltd had come to the same conclusion and this book is our attempt to fulfil that need.

It has proved to be a daunting task, for the ever widening and constantly changing scope of food law has made it difficult to decide what to include and what to omit. Another difficulty has been that there will be further important developments during the next year or two as the EC Single Market is established.

My approach has been to draw an outline portrait of food law and then to concentrate on important and long-established provisions which have been re-enacted in the Food Safety Act 1990, to explain the new provisions of that Act, and to clarify the role and influence of EC law. I have been influenced in my choice of topics by my experience as a consultant which suggests the issues which continue to give rise to difficulties of interpretation and implementation.

Another problem facing the author of a book on food law is that so much of it is new. The Food Safety Act 1990 is, of course, the most portentous source of new law and many EC developments also break entirely new ground. With only a limited period of operation, and without the benefit of judicial interpretation, I have felt it necessary to be cautious in offering opinions on new law, and I have tried to make my summaries of it as accurate as possible. However, the fact remains that this is a small book dealing with a huge subject and I have been obliged to seek a balance between essential principles and detail.

Food law is a fascinating subject. It has an infinite capacity to surprise; it poses the most intractable of problems; it attempts to control an ever changing and highly technological industry; and it can be most rewarding. If this modest book stimulates an interest in the subject I will be well satisfied.

Finally, I freely and gratefully acknowledge the help I have received and the collective wisdom I have enjoyed from trading standards and environmental health officers, from lawyers I have worked with over the years, from my friends in food science and technology and from my colleagues in Butterworth Law Publishers Ltd.

AA Painter
Aldwick, West Sussex
January 1992

Contents

Chronological Table of Statutes

References in this Table to *Statutes* are to Halsbury's Statutes of England (Fourth Edition) showing the volume and page at which the annotated text of an Act may be found.

Chronological Table of Statutory Instruments

SI—*contd*		PARA
1989/533	Preservatives in Food Regulations 1989	
	reg 7	3.1.2
1147	Water Supply (Water Quality) Regulations 1989	4.5.2
2061	Bovine Offal (Prohibition) Regulations 1989	11.2.2; 14.4.2, 14.6.5
2383	Milk (Special Designations) Regulations 1989	9.2.9; 11.2.2
1990/607	Milk and Milk Products (Protection of Designations) Regulations 1990	5.3.2; 7.1.3
998	Scotch Whisky Order 1990	5.3.2
1084	Preserved Sardines (Marketing Standards) Regulations 1990	7.4.1
1179	Spirits Drinks Regulations 1990	5.3.2; 6.4.21; 7.1.10
1323	Ungraded Eggs (Hygiene) Regulations 1990	11.2.2
1363	Common Agricultural Policy (Wine) Regulations 1990	7.2.1
1383	Food Safety Act 1990 (Commencement No 1) Order 1990	4.1.1
1431	Food Hygiene (Amendment) Regulations 1990	6.4.15; 13.1.2
2372	Food Safety Act 1990 (Commencement No 2) Order 1990	4.1.1
2462	Food Safety (Enforcement Authority) (England and Wales) Order 1990	2.4.3; 17.1.1
2463	Food Safety (Sampling and Qualifications) Regulations 1990	2.5.2; 17.1.1, 17.1.6, 17.3.3, 17.3.4, 17.4.3, 17.6.1
	reg 3	2.6.1
	5(1)	2.6.2
	7	2.7.2
2486	Food Safety Act 1990 (Consequential Modifications) (England and Wales) Order 1990	4.1.5
2487	Food Safety Act 1990 (Consequential Modifications) (No 2) (Great Britain) Order 1990	4.1.5
2490	Food (Control of Irradiation) Regulations 1990	9.2.9
2614	Detention of Food (Prescribed Forms) Regulations 1990	10.2.2; 14.2.4
2615	Quick-frozen Foodstuffs Regulations 1990	13.1.3, 13.3.1
1991/100	Food Safety (Improvement and Prohibition – Prescribed Forms) Regulations 1991	11.2.4, 11.4.2, 11.4.5, 11.4.7
370	Imported Food (Peruvian Foodstuffs) Regulations 1991	14.6.2
762	Food Safety (Northern Ireland) Order 1991	4.1.2
1343	Food Hygiene (Amendment) Regulations 1991	13.1.2
1382	Price Marking Order 1991	16.5.1
	art 4	16.5.4
	(3)(i)	16.5.8
	7	16.5.6
	8	16.5.8
1476	Food Safety (Exports) Regulations 1991	4.3.2; 17.1.1, 17.1.2
1593	Hormonal Substances (Food Sources) (Animals) Regulations 1991:	14.6.3
1958	Food Protection (Emergency Prohibitions) (Paralytic Shellfish Poisoning) (No 10) Order 1991	14.6.4
2246	Bovine Spongiform Encephalopathy Order 1991	14.6.5
2486	Imported Food and Feedingstuffs (Safeguards against Cholera) Regulations 1991	14.6.2
2825	Food Premises (Registration) Regulations 1991	12.1.3

Table of European Communities Legislation

Treaties

Secondary Legislation

DIRECTIVES

REGULATIONS

Table of Cases

Table of Abbreviations

AC	Appeal Cases Law Reports
All ER	All England Law Reports
Art	Article
CA	Court of Appeal
CCA	Court of Criminal Appeal
CMLR	Common Market Law Reports
Codex	Codex Alimentarius Commission
COM	Command (Royal)
Commrs	Commissioners (Customs and Excise)
Cr App R	Criminal Appeal Reports
Crim LJ	Criminal Law Journal
Crim LR	Criminal Law Review
DC	Divisional Court
Div	Division
DoH	Department of Health
EC	European Community
ECR	European Court Reports
Edn	Edition
EEC	European Economic Community
FAC	Food Advisory Committee
FSC	Food Standards Committee
HL	House of Lords
HMSO	Her Majesty's Stationery Office
IFST	Institute of Food Science & Technology
IR	Irish Law Reports
J and JJ	Justice, Justices
JC	Justiciary Cases
JP	Justice of the Peace Reports
JP Jo	Justice of the Peace Journal

KB	King's Bench Law Reports
LCJ	Lord Chief Justice
LGR	Local Government Reports
LJ and LJJ	Lord Justice, Lord Justices
LJR	Law Journal Reports
LR	Law Reports
LS Gaz	Law Society's Gazette
LT	Law Times
LTJ	Law Times Journal
MAFF	Ministry of Agriculture, Fisheries and Food
OJ	Official Journal of the European Communities
Paras	Paragraphs
QB	Queen's Bench Law Reports
r	rule
reg	regulation
RPC	Report of Patent, Design and Trade Mark Cases
SI 1991/2651	Statutory Instrument of 1991 Number 2651
s	section of a statute
Sch	Schedule
Tr L	Trading Law
WLR	Weekly Law Reports

Chapter 1

The Development of Modern Food Law

1.1 The Origins of Food Law

1.1.1 Food legislation may be traced back to the Middle Ages and the development of modern food law has been an evolutionary process. As government perceived a need for further measures on food, Acts of Parliament were passed and, in later years, a multiplicity of regulations was made to control the manufacture, safety, packaging and description of food. However, since the accession of the United Kingdom to the Treaty of Rome in 1972 there has been a gradual harmonisation of United Kingdom food law with legislation of the Community. Today, most United Kingdom food law flows from the implementation of EC Directives and Regulations.

1.1.2 The Food Safety Act 1990 which, from 1 January 1991, repealed almost all primary criminal law on food control, is not directly derived from EC law but is compatible with all existing EC requirements. Directives on such matters as basic food safety and enforcement are given expression in the Act but, more importantly in the EC context, ministers are empowered by the Act to make orders and regulations to implement future EC legislation and to amend existing primary and secondary legislation for that purpose.

1.2 The Nature of European Community Food Law

For a proper understanding of food law it is helpful to appreciate the nature and effect of EC legislation. Regulatory instruments of the EC are of four kinds:

1.2.1 *A directive* is a requirement by the Council of the EC that all member states shall adopt national measures to meet its objectives. It is a statement of Community objective which must be given effect in every member state by national legislation or, where permitted, by administrative or voluntary means. A time limit for implementation by member states is normally laid down in each directive and failure to implement a directive in full, or by the time due, is actionable by the EC Commission before the European Court of Justice. Directives have been made to cover a wide range of matters including labelling and advertising, additives, specific product categories, residues in food, enforcement and product liability. In the United Kingdom, directives are implemented by the Food Safety Act 1990 or regulations made under or continued in force by that Act and, in relation to product liability, by Part I of the Consumer Protection Act 1987.

1.2.2 *A regulation* is directly applicable to and binding on all member states. It is necessary to give effect to such regulations in the United Kingdom by legislation imposing penalties for non-observance and enforcement. Regulations cover such matters as marketing standards for milk, eggs, fruit and vegtables, wine, poultrymeat and spirit drinks. There has been a tendency in recent years for the greater use of regulations rather than directives because of the tardy implementation of directives by certain member states.

1.2.3 *A decision* is issued by the Council of Ministers or the Commission to any government, organisation or individual and is binding on those named. Decisions are sometimes used by the Council of Ministers to restrain a member state from making new national law in advance of EC consideration of and legislation on the issue concerned.

1.2.4 *A communication, recommendation or opinion* is not binding but expresses the Council's or Commission's view on policy. It is usually concerned with matters of broad policy but is sometimes issued in relation to analysis of food, the documentation to be used in food control, the practical effects of judgments of the European Court of Justice, or other related matters.

1.3 The Making of EC Law

1.3.1 There are three bodies involved in the EC law-making process. They are the Council of Ministers, the Commission and the European Parliament. The right of initiative lies with the Commission only and the Commission remains the guide, indeed the master, of a proposal throughout the law-making procedure. Before the adoption of the Single European Act[1] it was necessary to achieve unanimity before the proposal could be submitted to the Council of Ministers for adoption. It was often the case that one or more member states would veto a proposal, thus delaying agreement. Some important measures took upwards of seven years between proposal and adoption and were frequently in an emasculated state in their final form. The role of the European Parliament was peripheral and of little consequence.

1 See the European Communities (Amendment) Act 1986.

1.3.2 The European Single Act introduced a system of majority voting under which the Commission may choose between three different systems of consultation with member states. These are:

Advisory Committee Procedure. This consists of representatives of member states and is chaired by the Commission. The Commission must take account of but need not act on the views of the member states. It is not surprising that this is the procedure most favoured by the Commission.

Management Committee Procedure. This also consists of representatives of the member states and is chaired by the Commission, but qualified majority voting applies. If a qualified majority (see below) is in favour then the proposal is adopted. If there is neither a qualified majority nor a blocking minority against, the

Commission is free to adopt the proposal. This is known as the 'no opinion' procedure.

If there is a negative vote by a blocking minority the Commission may drop the proposal or produce a compromise text. If there is then no agreement to the original proposal or the compromise one, but the Commission believes the matter to be important, it may be referred to the Council of Ministers.

Regulatory Committee Procedure. The composition is as above and qualified majority voting applies. If a qualified majority is in favour of a proposal it is adopted. If there is no qualified majority and no blocking minority, the matter is referred to the Council of Ministers. If the Council takes no decision, the Commission proposal prevails. This is known as the 'safety net procedure'. There is also what is known as a 'double safety net procedure' by which in circumstances where there is no qualified majority in favour or blocking minority against a proposal, the matter is referred to the Council of Ministers but the Council is obliged to adopt the same voting procedure as the Regulatory Committee, thereby forcing the Commission to develop a compromise proposal which will find a qualified majority in both the Committee and the Council.

1.3.3 The voting power of each member state is as follows:

	Votes
UK/Germany/Italy/France	10
Spain	8
Netherlands/Belgium/Greece/Portugal	5
Denmark/Ireland	3
Luxembourg	2
	76

The votes needed for the approval or blocking of any proposal are:

Total number of votes	76
Qualified majority	54
Blocking minority	23

1.3.4 To British eyes these procedures are quaint, cumbersome and undemocratic but it should be borne in mind that the institutions of the EC are still evolving and by general agreement require further reform to make them more accountable for their decisions and actions. In matters of food law it has been clear for some time that the internal market is governed by Brussels, and national governments in member states have less and less autonomy in controlling their own internal markets.

1.3.5 From the adoption of the Single European Act[1] a so-called 'cooperation procedure' was adopted. This has created a stronger European Parliament (EP) and a less powerful Council of Ministers. The Commission is now in a strong position holding the centre ground between Parliament and the Council of Ministers. The debate about European federalism and monetary union may result in significant changes in the roles of the Commission, the Council and the EP but the following illustration of current procedures may be helpful.

1 See the European Communities (Amendment) Act 1986.

Cooperation procedure under the Single European Act

Commission Services Technical Preparation

Commission Proposal

↓

First EP reading: adoption of Opinion (no time limit)

Opinion of the Economic and Social Committee if required

↓

First Council reading: adoption of common position by qualified majority – no time limit

↓

Second EP reading – maximum of three months*

↓

Approval of Council's position or no decision

Rejection of Council's position

↓

Proposal for amendments (absolute majority)

↓

Council adopts the Act definitively (no time limit)

Council may adopt the Act by unanimity and within three months*

↓

Commission selects among amendments proposed and transmits to Council within one month revised proposal (and separately amendments not incorporated in the new text)

↓

Council adopts within three months* Commission's revised proposal (qualified majority)

Council amends within three months* Commission's revised proposal, eg adopts EP amendment rejected by Commission (unanimity)

↓

No Council decision within three months – deemed not to have been adopted.

* This period may be extended by a maximum of one month by common accord between Council and the EP

1.3.6 The effect of the Single European Act has been to speed up the process of harmonisation of EC food law. In principle this is beneficial but it has resulted in some hasty, ill-considered and very badly drafted legislation. However, it remains desirable that all EC food law should be harmonised by 1 January 1993 (the date on which the Single European Act takes effect) to achieve the basic objective of the Act, namely the elimination of all legislative barriers to trade between member states.

1.4 Mutual Recognition of Products

1.4.1 The celebrated Cassis de Dijon judgment of the European Court of Justice (ECJ) of 1979[1] firmly established the principle enshrined in Article 30 of the Treaty of Rome that any product which could be lawfully marketed in one member state must be admitted to the markets of other member states unless it could be excluded on grounds of public health under Article 36 of the Treaty. Numerous judgments of the ECJ in the years from 1979 to 1992 have confirmed the principle and restricted the freedom of member states to exclude products under Article 36.

1 *Rewe-Zentral AG v Bundesmonopolverwaltung für Branntwein* [1979] ECR 649, [1979] 3 CMLR 494, ECJ.

1.4.2 Although the Single European Act is essentially an expression of political intent, the ECJ has supported the principle of mutual recognition of products with vigour.[1] The effect has been to create market conditions within the EC for food products in accordance with the Single Market objectives well in advance of 1 January 1993.

1 For a discussion of ECJ judgments, see *Butterworths Law of Food and Drugs*.

1.5 Non-observance of Community Law

1.5.1 Failure of a member state to implement EC law within its territory in full or by the due date, subject to such derogations as may be agreed, is actionable before the ECJ. The EC Commission has proved itself to be diligent in bringing such cases before the Court and has a long record of success in such actions. Similarly, referrals to the ECJ by national courts of cases involving disagreement on the interpretation of Community law or the non-observance of such law by individuals and companies have resulted in consistency of action and interpretation.

1.5.2 However, actions referred to the ECJ inevitably take a very long time – up to four or five years in many cases. Conscious of the need for swifter resolution of such problems the Commission has introduced a procedure under which it will seek to resolve such matters by negotiation.[1] Many problems for food manufacturers, importers and retailers result from uneven interpretation of food law by the enforcement authorities of member states which, despite the work being done on coordinating enforcement throughout the Community (see Chapter 2) will remain a problem for some time to come by virtue of the very diverse and fragmented arrangements for enforcement in certain member states.

1 *Butterworths Law of Food and Drugs.*

1.5.3 In 1988 the Community established a Court of First Instance[1] which initially may hear actions brought by natural and legal persons against the Commission and related matters. The powers of this court may be extended to other cases in due course.

1 OJ No L 319; 25.11.88 and OJ No C 136; 05.06.90.

1.6 EC Food Law Policy

1.6.1 From time to time the EC Commission issues Communications on food law policy which give guidance on forward programmes for food legislation and reviews existing legislation. Such documents are useful to those seeking long-term guidance on EC intentions.[1]

1 OJ No C 271; 24.10.89.

1.7 Negotiation on EC Food Law

1.7.1 EC member states are free to legislate nationally only on matters which are not currently the subject of EC law. Even then, proposals for legislation have to be submitted to the Commission before enactment. The scope for purely national legislation is thus severely limited. The Food Safety Act 1990 requires ministers to consult with organisations representing interests likely to be affected by proposed legislation.[1] The Act makes no distinction between EC and national legislation and government departments are generally diligent in fulfilling this obligation and issue detailed consultative papers on most proposals to consumer, enforcement and trade interests.

1 Section 48(4) of the Food Safety Act 1990.

1.7.2 It is important, however, to appreciate that in submitting their views to government, interested parties are essentially negotiating with the negotiators and not with the legislators. Nevertheless, ministers and officials of the Ministry of Agriculture, Fisheries and Food and the Department of Health have proved themselves willing and able to comprehend the views submitted to them and skilful in representing those views in Brussels.

1.7.3 Most sectors of the food industry and consumer interests have established trade representative bodies in Brussels, or thereabouts, to argue the case for their members directly with the Commission. Some, such as those representing consumers, retailers, the meat and dairy industries have become influential but others have been less successful.

1.8 United Kingdom Implementation of EC Food Law

1.8.1 Despite its unjustified reputation as the odd-man-out in Europe, the United Kingdom has a good record for implementing EC directives and giving the force of law to EC regulations by the due time and in every detail. Implementation is generally by way of the Food Safety Act 1990, the Weights and Measures Act 1985 and the Consumer Protection Act 1987 and regulations and orders made or continued in force under those statutes.

1.8.2 The food manufacturing and distribution industries of the United Kingdom have to comply with a formidable framework of legislation.[1] Although much of the law currently in force has been derived from legislation dating back up to one hundred years or more, there has been a general shift of emphasis away from precise and very detailed law towards general statutory duties and codes of practice. Examples are the food safety requirement[2] and codes of practice issued by the ministers for the guidance of food authorities.[3]

1 See Appendix 1.
2 Food Safety Act 1990, s 8.
3 Food Safety Act 1990, s 40.

1.8.3 For a list of the principal EC measures of food law, see Appendix B.

Chapter 2

Central and Local Government Control of Food

2.1 Introduction

From the earliest times the creation and administration of food law in the United Kingdom has, with a few exceptions, been the responsibility of two or more departments of central government and a variety of local authorities. Today, despite the revolution in food law emanating from our membership of the European Community and the enactment of the Food Safety Act 1990, central government responsibilities are still shared between the Ministry of Agriculture, Fisheries and Food and the Department of Health, and the bulk of enforcement work is in the hands of county councils and district councils.

Within central government, policy, law making and some aspects of enforcement are the responsibility of ministers and their administrative and legal staff and specialist inspectorates concerned with horticultural marketing, pesticide and veterinary residues, meat and milk. Within local government enforcement work is allocated to public analysts, environmental health officers, trading standards officers, veterinarians and their subordinates.

Despite the creation of elaborate liaison and coordination arrangements there is an inevitable loss of efficiency and immediacy and a significant waste of public resources. Competition for resources between authorities and departments, professional rivalries, fragmented training of staff and poor communication all militate against efficient administration.

Although ministers were well aware of the need for simplified and more efficient arrangements when drafting the Food Safety Bill, they were faced with constitutional, structural, political and economic difficulties which, at the time, made it difficult to create a single Ministry for Food and national or regional enforcement authorities. Certain changes within central government, such as the establishment of the Food Safety Directorate within the Ministry of Agriculture, Fisheries and Food and greater statutory powers for the coordination of local authority work have improved enforcement but the system remains much the same as that which has existed since the turn of the century. It is all profoundly unsatisfactory.

2.2 Ministerial Responsibilities

For the purposes of the Food Safety Act 1990 certain responsibilities are given to 'the Minister' which means the Minister of Agriculture, Fisheries and Food or the Secretary of State and in relation to Scotland, the Secretary of State. In relation to emergency control orders, see Chapter 11, the 'Minister' means the Minister of Agriculture, Fisheries and Food or the Secretary of State. Other responsibilities are given to ministers acting jointly, and the term 'the Ministers' means, in relation to England and Wales, the Minister of Agriculture, Fisheries and Food and the Secretaries of State respectively concerned with health in England and food and health in Wales and in relation to Scotland, the Secretary of State.[1]

1 Food Safety Act 1990, s 4.

For the purposes of the Food and Environment Protection Act 1985 'the Ministers' are defined as being the Minister of Agriculture, Fisheries and Food and the Secretary of State.

2.2.1 The functions and powers of the Minister are:

(i) to specify which functions under the Act shall be exercisable solely by one class of food authority;[1]
(ii) to specify which authorities shall enforce and execute regulations;[2]
(iii) power to take over the conduct of any proceedings which have been instituted by another person;[3]
(iv) power to issue, vary or give directions in relation to emergency control orders;[4]
(v) to approve apparatus etc for the cleansing of shell fish;[5]
(vi) power to direct that expenses incurred by county councils in the enforcement or execution of any provision of the Act be defrayed as expenses for special county purposes;[6]
(vii) power to apply, with modifications and adaptions, any other enactment to regulations or orders made under the Act;[7]
(viii) power to certify that the powers of entry conferred by s 32 of that Act should not be exercisable in relation to Crown premises.[8]

1 Food Safety Act 1990, s 5 (4).
2 Food Safety Act 1990, s 6 (4).
3 Food Safety Act 1990, s 6 (5).
4 Food Safety Act 1990, s 13.
5 Food Safety Act 1990, s 24 (3).
6 Food Safety Act 1990, s 46 (2).
7 Food Safety Act 1990, s 48 (1).
8 Food Safety Act 1990, s 54 (4).

2.2.2 The functions and powers of the Ministers are:

(i) power to provide exemptions from the definition of 'food';[1]
(ii) power to extend the definition of 'premises' to any ship or aircraft as specified;[2]
(iii) power to extend the meaning of 'sale';[3]
(iv) power to provide by order that functions under the Act be exercised by a particular food authority;[4]
(v) power to prescribe qualifications for authorised officers;[5];
(vi) power to direct that any duty imposed on food authorities shall be discharged by the Minister or the Ministers;[6]
(vii) requirement to make regulations as to health, safety and hygiene in relation to food;[7]

(viii) requirement to make regulations for the enforcement of Community provisions;[8]
(ix) power to make regulations as to commercial operations with respect to novel foods, food sources and genetically modified food sources and prohibiting the importation of food of a class specified;[9]
(x) power to make regulations as to the licensing of producers and sellers of milk and the special designations of milk;[10]
(xi) power to make regulations requiring the registration or licensing of food premises;[11]
(xii) power to require by order persons carrying on businesses of a specified class or description to afford facilities for the taking of samples or to furnish information as required;[12]
(xiii) power to make regulations prohibiting or regulating the carrying out of certain commercial operations with respect to food, food sources or contact materials;[13]
(xiv) power to make regulations requiring persons carrying on any activity to which regulations apply to keep records or registers and allow inspection by the public thereof, regulations laying down the conditions subject to which licences may be issued and to provide for appeals in relation thereto;[14]
(xv) power to provide for the qualifications of public analysts by regulations;[15]
(xvi) power to prescribe by regulations arrangements made between a public analyst and a food business in relation to his being engaged directly or indirectly in that business;[16]
(xvii) power to provide for the qualifications of food examiners or food analysts by regulations;[17]
(xviii) power to make regulations as to sampling and analysis;[18]
(xix) power to issue codes of practice for the guidance of food authorities;[19]
(xx) power to issue directions to food authorities as to compliance with a code of practice;[20]
(xxi) power to require the submission of reports and returns by a food authority;[21]
(xxii) power to order another food authority to discharge the duties of a food authority in default;[22]
(xxiii) power to cause a local inquiry to be held in respect of (xxii) above;[23]
(xxiv) power to recover expenses from a food authority in default;[24]
(xxv) power to make regulations authorising charges to be imposed by food authorities;[25]
(xxvi) to determine the remuneration and allowances to be paid to the chairmen of any tribunal constituted under regulations or the Act;[26]
(xxvii) power to apply with modifications or adaptions, other enactments to regulations or orders made under this Act;[27]
(xxviii) power to prescribe the form of documents;[28]
(xxix) power to make emergency orders where there is a hazard to human health through the consumption of food;[29]
(xxx) power to consent to the doing of things which would otherwise contravene an emergency order and to give directions in relation to an emergency order;[30]
(xxxi) power to authorise investigating officers;[31]
(xxxii) power to make regulations for the control and use of pesticides;[32]
(xxxiii) power to make codes of practice as to the use etc of pesticides;[33]
(xxxiv) power to approve pesticides for use and make charges therefor.[34]

1 Food Safety Act 1990, s 1 (2) (d).
2 Food Safety Act 1990, s 1 (3).
3 Food Safety Act 1990, s 2 (1) (b).
4 Food Safety Act 1990, s 5 (4).
5 Food Safety Act 1990, s 5 (6).
6 Food Safety Act 1990, s 6 (3).
7 Food Safety Act 1990, s 16.
8 Food Safety Act 1990, s 17.
9 Food Safety Act 1990, s 18 (1).
10 Food Safety Act 1990, s 18 (2).
11 Food Safety Act 1990, s 19 (1).
12 Food Safety Act 1990, s 25 (1).
13 Food Safety Act 1990, s 26 (1).
14 Food Safety Act 1990, s 26 (2).
15 Food Safety Act 1990, s 27 (2).
16 Food Safety Act 1990, s 27 (5).
17 Food Safety Act 1990, s 30 (9).
18 Food Safety Act 1990, s 31.
19 Food Safety Act 1990, s 40 (1).
20 Food Safety Act 1990, s 40 (2).
21 Food Safety Act 1990, s 41.
22 Food Safety Act 1990, s 42 (1).
23 Food Safety Act 1990, s 42 (2).
24 Food Safety Act 1990, s 42 (4).
25 Food Safety Act 1990, s 45.
26 Food Safety Act 1990, s 47.
27 Food Safety Act 1990, s 48 (1).
28 Food Safety Act 1990, s 49 (1).
29 Food & Environment Protection Act 1985, s 1.
30 Food & Environment Protection Act 1985, s 2.
31 Food & Environment Protection Act 1985, s 3.
32 Food & Environment Protection Act 1985, s 16.
33 Food & Environment Protection Act 1985, s 17.
34 Food & Environment Protection Act 1985, ss 16 & 18.

2.3 Advisory Committees

A number of committees have been appointed by the Ministers to advise them on matters of food quality, description and safety. Their work ensures that the Ministers have access to high levels of scientific and technical expertise in order that they may properly discharge their duties under the Food Safety Act 1990 and the Food and Environment Protection Act 1985. The current committees are:

2.3.1 *Food Advisory Committee.* This Committee is responsible for advice to the Ministers on the composition, labelling and advertising of food and the control of additives, contaminants and other substances which are, or may be present, in food used in its preparation. This Committee is the longest standing of all such committees and its reports (and those of its predecessors the Food Standards Committee and the Food Additives and Contaminants Committee) form an important source of reference material and are often used in the interpretation of food law.[1]

1 See Div D, *Butterworths Law of Food & Drugs.*

2.3.2 *Advisory Committee on Pesticides.* This Committee advises the Ministers on regulations for the control of pesticides, as to approval of pesticides which they contemplate giving, revoking or suspending and any conditions which should attach to approvals.

2.3.3 *Veterinary Products Committee.* This Committee advises the Ministers on safety, quality and efficacy in relation to veterinary use of any substance or article which is not an instrument, apparatus or application to which any provisions of the Medicines Act 1968 are applicable.

2.3.4 *Committee on Microbiological Safety of Food.* Advice on matters remitted to it by the Ministers is given by this Committee in relation to the microbiological safety of food and on such matters as it considers need investigation. Its main role is to look at specific questions relating to the increasing incidence of microbiological illnesses of food-borne origin, particularly from salmonella, listeria and campylobacter; to establish whether this is linked to changes in agriculture and food production, food technology and distribution, retailing, catering and food handling in the home; and to recommend action where appropriate.

2.3.5 *Steering Group on Food Surveillance.* This group keeps under review the possibilities of contamination of any part of the national food supply and reviews where necessary the intake of individual additives and nutrients and recommends to the Ministers responsible for food quality and safety the programme of work necessary to ensure that the food intake of the population is both safe and nutritious.

2.3.6 *Advisory Committee on Novel Foods and Processes.* Advice is given to the Ministers on any matters relating to the irradiation of food or to the manufacture of novel foods or foods produced by novel processes and having regard where appropriate to the views of relevant expert bodies.

2.3.7 *Committee on the Toxicity of Chemicals in Food, Consumer Products and the Environment.* This Committee assesses and advises on the toxic risk to humans of substances which are used or proposed to be used as food additives, or used in such a way that they might contaminate food through their use or natural occurrence in agriculture, including horticulture and veterinary practice or in the distribution, storage, preparation, processing or packaging of food.

2.3.8 *Committee on Carcinogenicity of Chemicals in Food, Consumer Products and the Environment.* This Committee assesses and advises on the carcinogenic risk to man of substances which are used or proposed to be used as food additives, or used in such a way that they might contaminate food through their use or natural occurrence in agriculture, including horticulture and veterinary practice or in the distribution, storage, preparation, processing or packaging of food.

In addition to the foregoing standing Committees, Ministers from time to time, appoint special working groups to consider particular issues or proposals.

2.4 Food Authorities

Definition

2.4.1 By s 5 (1) of the Food Safety Act 1990 food authorities in England and Wales are the London Boroughs, the metropolitan and non-metropolitan district councils and the non-metropolitan county councils, the Common Council of the City of London and the appropriate Treasurer of the Inner Temple or the Middle Temple. In Scotland the food authorities are the islands or district councils.

Local authorities may combine or make arrangements for the discharge of certain of their duties by other local authorities.[1] Following the abolition of the metropolitan county councils in England in 1985 functions in relation to food were, in certain areas, assigned to a single ad hoc authority for the former metropolitan county area.[2] Certain port health authorities in England and Wales[3] and port local authorities in Scotland[4] have had duties in relation to food assigned to them and are deemed to be food authorities.[5]

1 Local Government Act 1972, s 101.
2 Local Government Act 1985, Sch 8.
3 Public Health (Control of Disease) Act 1984, ss 2 & 7.
4 Public Health (Scotland) Act 1897, s 172.
5 Food Safety Act 1990, s 5 (3).

Statutory Duty

2.4.2 Every food authority is under a statutory obligation to enforce and execute the provisions of the Food Safety Act 1990[1] within their area except where that duty is expressly or by necessary implication imposed on some other authority. To distinguish between the broad concept of a 'food authority' and an authority which is given express responsibility for the enforcement of particular provisions of the Act or regulations made or continued in force thereunder, the term 'enforcement authority' has been created.[2] Thus a county council is a food authority for the whole of its area but an enforcement authority only for the purposes expressly allocated to it by the Act or an order made under it. On the other hand, a district council within that county is a food authority for the whole of its area but an enforcement authority only for such provisions of the Act as have been allocated to it. Only authorised officers of an enforcement authority enjoy the powers of entry, investigation and sampling conferred by the Act or regulations.

1 Food Safety Act 1990, s 6 (2).
2 Food Safety Act 1990, s 6 (1).

Allocation of Duties

2.4.3 By virtue of the Food Safety Act 1990 district councils within the metropolitan areas of England and Wales and islands and district councils in Scotland enjoy the status of unitary authorities and thus responsibility for the enforcement of all provisions of the Food Safety Act 1990 for which food authorities are responsible. In the shire counties of England and Wales the county councils and the district councils enjoy concurrent powers of enforcment of the Act except where, by order, the Minister has decreed otherwise. By the Food Safety (Enforcement Authority) (England and Wales) Order 1990/2462 the provisions in respect of emergency prohibition notices and Orders[1] may be exercised only by the district councils in such areas and the provisions as to false presentation and description of food[2] may be enforced solely by the county councils.

Further clarification as to which authority in the shire counties should carry out certain functions may be found in Statutory Code of Practice No 1: Responsibility for Enforcement of the Food Safety Act 1990.[3] The Code stresses the need for adequate liaison arrangements between district and county councils to ensure full cooperation in enforcement and the passing of information between authorities. It further recommends that district councils should investigate and take legal proceedings in all cases of contamination by micro-organisms or their toxins such as salmonella, listeria or botulism and all cases of mould or foreign matter in or on food. County councils should undertake routine checks and analysis of food for chemical contamination and improper use of additives and should normally investigate and take legal proceedings in cases of contamination by chemicals, subject to proper liaison with the district councils where there are public health implications. County councils should also investigate and take legal proceedings in cases of compositional offences, adulteration and misleading claims.

The appropriate tier of food authority for the enforcement of regulations made under the 1990 Act is decreed in the regulations themselves but the division of responsibility adheres, generally speaking, to the principles set out above.

1 Food Safety Act 1990, s 12.
2 Food Safety Act 1990, s 15.
3 HMSO, ISBN 0 11 321354 9.

Default of Food Authorities

2.4.4 If a local authority is in breach of a statutory duty *mandamus* (an order to command from the High Court) may lie. However, the Food Safety Act 1990 empowers the Ministers to take action against any such authority.

Where the Minister is satisfied that a food authority has failed to discharge any duty imposed on it by the Act and the general interests of consumers are affected by such default he may by order empower another food authority to discharge that duty. He may, but is not obliged to, cause a local inquiry to be held where he believes that a food authority is in default.[1]

Under former legislation the Minister was obliged to hold a local inquiry before any action could be taken against an authority in default. As far as is known there has never been such an inquiry despite the fact that a few authorities have paid little regard to their statutory responsibilities for food law enforcement. This may have had something to do with a lack of evidence of default and quite a lot to do with the indolence of Ministers of successive governments over the years. Ministerial statements since the enactment of the Food Safety Act 1990 suggest that there is now a new determination to ensure that the Act is fully and comprehensively enforced.

As an alternative to the powers described above the Ministers have the power to take over any duty imposed on food authorities.[2]

In order to monitor the activity of food authorities the Ministers are empowered to require reports and returns from food authorities.[3]

1 Food Safety Act 1990, s 42 (1).
2 Food Safety Act 1990, s 6 (3).
3 Food Safety Act 1990, s 41.

Codes of Practice

2.4.5 Ministers are empowered to issue codes of practice for the guidance of food authorities as regards the execution and enforcement of the Act or regulations or orders made under it.[1] Food authorities are required to have regard to the relevant provision of any such code and shall comply with any direction which is given to them by the Ministers to take any specifed steps to comply with a code.[2] Any such direction is enforceable by *mandamus* or, in Scotland, by an Order of the Court of Session.[3]

1 Food Safety Act 1990, s 40 (1).
2 Food Safety Act 1990, s 40 (2).
3 Court of Session Act 1988, s 45.

2.5 Authorised Officers

2.5.1 Fulfilment of the statutory duty to enforce the Food Safety Act 1990 and regulations or orders made thereunder is by way of 'authorised officers'. An authorised officer of a food authority means any person, whether or not an officer of the authority, who is authorised by them in writing, either generally or specially, to act in matters arising under the Act.[1]

Generally, authorised officers of district councils are environmental health officers and those of county councils are trading standards officers. Veterinarians and other appropriately qualified officers are employed by district councils to carry out inspections in accordance with EC requirements as to trade in meat and for the inspection of slaughterhouses and meat cutting premises.

1 Food Safety Act 1990, s 5 (6).

2.5.2 Ministers are empowered to make regulations prescribing the qualifications of authorised officers.[1] Only public analysts, food examiners and officers employed for the purpose of the examination and seizure of meat are subject to such regulations at present.[2] In order to comply with the EC Directive on the Official Control of Foodstuffs[3] work is being done within the EC Commission and in member states to improve the training and qualifications of authorised officers throughout the Community.

Both enviromental health officers and trading standards officers are trained in food legislation and enforcement techniques although their qualifications are not, for this purpose at least, subject to statutory requirement. There is, however, no doubt that the quality of food law enforcement officers in the United Kingdom is at least as high as the best in other member states.

1 Food Safety Act 1990, s 5 (6).
2 Food Safety (Sampling and Qualifications) Regulations 1990/2463; Authorised Officers (Meat Inspection) Regulations 1987/133.
3 Directive 89/397/EEC.

2.5.3 An authorised officer of a food authority is not personally liable for any act done by him within the scope of his employment and for the purpose of the execution or purported execution of the Act. Further, if an action is brought against an officer of a food authority in respect of an act done by him his employing food authority may indemnify him in whole or in part against any damages or costs which he may be ordered to pay.

There is a growing trend for food authorities to secure the services of consultants or freelance enforcement staff for the enforcement of the Act. This is authorised under the Act.[1] Such officers are not protected as are officers of the authority itself and the food authority is not empowered to indemnify them.

1 Food Safety Act 1990, s 5 (6).

2.6 Public Analysts

2.6.1 The public analyst is the key figure in the efficient enforcement of food law. He is the principal scientific adviser to the food authority and the success of food law enforcement depends on his scientific and interpretary skills.

Food authorities are obliged to appoint a public analyst[1] who must be qualified in accordance with regulations made by the Ministers.[2] The authority must pay to the public analyst such remuneration as may be agreed and may also appoint a deputy public analyst. The public analyst is deemed to be an officer of the food authority whether or not he is an employee of that authority.[3]

1 Food Safety Act 1990, s 27 (1).
2 Food Safety (Sampling and Qualifications) Regulations 1990/2463, reg 3.
3 Food Safety Act 1990, s 44 (4).

2.6.2 Some public analysts are officers of a food authority and work from laboratories provided by that authority. Others are independent. Many public analysts are appointed by one or more food authorities. No person may act as a public analyst for an area if he is engaged directly or indirectly in any food business in that area[1] and by regulations the Ministers have provided that no director, owner or employee of a food business, or partner in a food business, shall act as a public analyst for the area in which that business is situated.[2] Breach of this prohibition would not give rise to criminal liability but would invalidate the public analyst's appointment and would render any evidence he wished to give inadmissible.

1 Food Safety Act 1990, s 27 (2).
2 Food Safety (Sampling and Qualifications) Regulations 1990/2463, reg 5 (1).

2.7 Food Examiners and Food Analysts

2.7.1 The Food Safety Act 1990 introduced the concept of a food examiner and a food analyst. Food examination means microbiological examination[1] and a food examiner is defined as being a person who possesses the requisite qualifications to carry out examinations for the purposes of the Act.[2] Food authorities are not obliged to appoint a food examiner but most public analysts are also qualified as food examiners or employ persons who are so qualified.

1 Food Safety Act 1990, s 28 (2).
2 Food Safety Act 1990, s 30 (9).

2.7.2 A food analyst is a person having the same qualifications as a public analyst[1] but who has not been appointed as a public analyst. Food analysts provide analytical services to industry and a certificate of analysis given by a food analyst has evidential status (see **17.3.5**).

1 Food Safety Act 1990, s 30 (9).

2.8 The Government Chemist

The Government Chemist is the final arbiter for the purposes of the Food Safety Act 1990 as to the analysis of samples. An authorised officer who has retained part of a sample is required to submit it to the Government Chemist, or such other food analyst as the Government Chemist may direct, for analysis if he and the owner agree or if a court so orders.[1]

1 Food Safety (Sampling and Qualifications) Regulations 1990/2463, reg 7.

Chapter 3

Consumer Remedies

3.1 Introduction

3.1.1 Before considering the criminal law on food control it is important to have regard to civil law remedies for consumers who buy food which is of poor quality, misdescribed, unsafe, or defective in some other way. In certain respects civil and criminal law on food is complementary. A consumer who is aggrieved by a food product may decide to report the matter to a food authority for investigation and, if appropriate, prosecution under the Food Safety Act 1990. If the prosecution were successful it would then be open to the prosecutor to seek a compensation order from the criminal court in favour of the purchaser of the offending food.[1] The consumer may, on the other hand, decide to bring an action against the retailer or supplier for damages instead of or in addition to any criminal proceedings.

1 S 35 of the Powers of Criminal Courts Act 1973; s 40 of the Magistrates' Courts Act 1980; s 67 of the Criminal Justice Act 1982.

3.1.2 In establishing their food control and management systems food retailers and producers should also bear in mind the possible consequences of civil actions by consumers. Reasonable precautions and due diligence systems to avoid the commission of a criminal offence[1] will reduce the likelihood of defective products and thus the risk of civil actions. Warnings and instructions as to use to fulfil statutory requirements[2] may also serve to limit liability under the Consumer Protection Act 1987.[3]

1 S 21 of the Food Safety Act 1990; s 22 of the Food and Environment Protection Act 1985; s 34 of the Weights and Measures Act 1985.
2 Eg Food Labelling Regulations 1984, regs 12, 23 and 36; Preservatives in Food Regulations 1989, reg 7.
3 Consumer Protection Act 1987, s 3 (2).

3.1.3 There are three courses of action which an aggrieved consumer may use to obtain damages for a defective food product, namely breach of contract, tort or strict product liability.

3.2 Breach of Contract

3.2.1 For a contract to exist there must have been a pecuniary consideration, ie the food must have been bought and paid for. The extended meaning of 'sale' in s 2 of the Food Safety Act 1990 does not apply to the law of contract. Further, the contract must be complete before any action for breach of contract can take place. There must be an offer to buy and an acceptance of that offer by the vendor. Thus the selection of an item of food from a supermarket shelf does not establish a contract. The contract comes into being when the food is presented at the check-out and the money is handed over.[1]

1 *Pharmaceutical Society of Great Britain v Boots Cash Chemists (Southern) Ltd* [1952] 2 QB 795, [1952] 2 All ER 456.

3.2.2 Where a consumer has bought a food product which is defective he or she may sue the vendor for damages for breach of contract. Action may only be against the immediate vendor who is normally the retailer. A consumer has no case against the manufacturer or processor of a defective food product for breach of contract. A retailer against whom an award of damages has been or may be made can, subject to such exemption clauses as may be in the contract, sue his supplier, perhaps a wholesaler or manufacturer, right through the chain of supply to the party responsible for the defect.[1]

1 *Kasler and Cohen v Slavouski* [1928] 1 KB 78: *Dodd and Dodd v Wilson and McWilliam* [1946] 2 All ER 691.

3.2.3 Only the actual purchaser of the food is entitled to sue the vendor in contract.[1] Thus a mother who bought food for her children which made them ill may be able to sue the retailer for pecuniary loss of the value of the food but not for the injury suffered by the children. In such a case the children would have to sue in their own right.

1 *Donoghue v Stevenson* [1932] AC 562, HL.

3.2.4 No proof of of negligence is required for an action in contract but where such proof does exist the buyer may choose to take action in tort as an alternative.

3.2.5 In every contract for the sale of food by retail there are three implied terms of contract.[1] They are:

(1) *The food must correspond to its description.* Any description of food by word of mouth, notice, advertisement, label or any other means which is different from the actual food sold is actionable for breach of contract. The term 'sale of goods by description' has acquired an extended meaning. A sale may be by description even though the buyer has seen and examined the food and has selected it for himself.[2] Food cannot, by virtue of the nature of the trade and legislation on labelling, be sold simply as a specific object.[3]

Bearing in mind the current vogue for more and more information to be given to consumers the right to sue for non-conformity to description is a valuable protection for consumers.

(2) *The food must be of merchantable quality.* The term 'merchantable quality' has existed in contract law for many years and has been the source of much uncertainty. The statutory definition[4] now is –

> 'Goods of any kind are of merchantable quality . . . if they are as fit for the purpose or purposes for which goods of that kind are commonly bought as it is reasonable to expect having regard to any description applied to them, the price (if relevant) and all other relevant circumstances'.

It has long been established that this definition links merchantability to fitness for purpose rather than to commercial saleability.[5] This is particularly important in relation to sales of food products, especially those for special nutritional purposes.

Linked as it is to both description and fitness for purpose, actions on grounds of want of merchantable quality can include most defects in food products.

(3) *Food must be reasonably fit for a particular purpose.* Where a buyer makes it known that he or she is seeking food for a particular purpose and is relying on the skill or judgment of the vendor to meet that requirement then there is an implied term in every consumer contract that the food is reasonably fit for that purpose.[6]

Although this term may have some benefit for consumers, such as where a buyer asks a butcher for a cut or species of meat for a particular culinary purpose, it is not generally employed in food cases because of the wider application of the 'description' and 'merchantable quality' terms explained above.

1 Sale of Goods Act 1979, s 14.
2 *Grant v Australian Knitting Mills Ltd* [1936] AC 85, PC.
3 *Wallis v Russell* [1902] 2 IR 585, CA.
4 Sale of Goods Act 1979, s 14 (6).
5 *Cammell Laird & Co Ltd v Manganese Bronze and Brass Co Ltd* [1934] AC 402, HL.
6 Sale of Goods Act 1979, s 14 (3).

3.3 Liability in Tort

3.3.1 Until 1932 there was no general remedy in tort for damage caused by defective products even where the person responsible for the defect could be proved to have been negligent. However, in *Donoghue v Stevenson*[1] it was held that the supposed rule of non-liability was misconceived and that a manufacturer might be liable on proof of negligence when personal injury or property damage was suffered by a consumer of his product. In that case a young lady suffered gastro-enteritis as a result of drinking a bottle of ginger beer contaminated by the remains of a decomposed snail. The bottle had been bought from the retailer by her companion thus depriving her of action at contract. In his judgment, Lord Atkin said –

> 'My Lords . . . a manufacturer of products, which he sells in such a form as to show that he intends them to reach the ultimate consumer in the form in which they left him with no reasonable possibility of intermediate examination, and with the knowledge that the absence of reasonable care in the preparation or putting up

of the products will result in an injury to the consumer's life or property, owes a duty to the consumer to take reasonable care.'

1 *Donoghue v Stevenson* [1932] AC 562, HL.

3.3.2 Once it was established that there was a general duty of care for manufacturers the courts began to apply the rule broadly both in the range of products subject to the rule and the degree of negligence which had to be proved. However, there remained a need for a regime of strict product liability. In 1977 the Law Commission in its report 'Liability for Defective Products'[1] and the Royal Commission on Civil Liability and Compensation for Personal Injury (the Pearson Commission)[2] recommended that producers should be strictly liable for death or personal injury caused by a defect in their products. In 1985 Council Directive 85/374/EEC on the approximation of the laws, regulations and administrative provisions of the member states concerning liability for defective products was adopted. For the first time a uniforn regime of strict product liability was introduced throughout the European Community. The Directive was implemented in the United Kingdom from 1 March 1988 by Part 1 of the Consumer Protection Act 1987.

1 Cmnd 6381 – 1977.
2 Cmnd 7054 – 1978.

3.3.3 Although the changes in the law introduced by the Consumer Protection Act 1987 are of fundamental importance, the remedy of action at tort remains available to consumers. Bearing in mind, however, that it is not necessary to prove negligence to succeed in an action under that Act, it is unlikely that actions in tort will occur in the future in relation to food products.

3.4 Strict Product Liability

3.4.1 The purpose of Council Directive 85/374/EEC, implemented in the United Kingdom by Part 1 of the Consumer Protection Act 1987, is to reduce distortions in competition between member states and to give similar protection to consumers throughout the Community against defective products.

Strict product liability means that any person can sue for compensation for death, personal injury or damage to property valued at more than £275.[1] It is neither necessary for the complainant to prove negligence by the manufacturer nor that he or she has a contract with the retailer.[2]

1 Consumer Protection Act 1987, s 5 (4).
2 Consumer Protection Act 1987, s 2.

Burden of Proof

3.4.2 It is necessary for a plaintiff seeking damages under the Act to prove, on the balance of probability, that the death, injury or damage was wholly or partly attributable to a defect in the product. A person who suffered illness after eating a certain food would have to prove that the food was the cause of the illness by way of contamination or composition. It would not, it is submitted, be enough to show that an illness already existing before consumption was exacerbated by the food and it is unlikely that an illness caused by gluttony could be attributed to the food.

3.4.3 *Consumer Remedies*

The consumption of contaminated food and likely litigation arising therefrom have focussed attention on the nature of injury for which compensation may be sought. Actual physical injury is beyond doubt but what of emotional stress? The courts appear to be taking a sympathetic view of cases where mothers claim to have suffered mental trauma as a result of finding glass in jars of baby food even though the baby was not injured.

Food Products Subject to the Act

3.4.3 Generally speaking all consumer goods may be the source of an action under the Act. However, defects in game or agricultural produce are exempted if it has not undergone an industrial process.[1] There is much uncertainty about what an industrial process is in this context. Take, for example the humble cucumber. In its harvested form it is undoubtedly agricultural produce which is defined as produce of the soil, of stock-farming or of fisheries. But what if it is washed sliced and pre-packed? Has it undergone an industrial process? What too of fresh butchers' meat? A whole carcase is certainly the product of animal husbandry and thus exempt. But if it is cut up into joints, has it undergone an industrial process?

In the absence of any judicial interpretation there is no certain answer to these questions but s 1 of the Act begins with the words:

> 'This Part shall have effect for the purpose of making such provision as is necessary in order to comply with the product liability Directive and shall be construed accordingly'.

If one reads the Directive carefully, especially the preamble thereto, it is impossible to escape the conclusion that it is aimed at defects in products which are the result of bad design, poor manufacture and human failings generally. If that is right, then it is likely that the cucumber and the fresh meat are indeed agricultural produce and exempt. However, this may not be the case if, for example, the cumcumber contains an excess of pesticide residue or the meat contains hormonal residues. The application of the pesticides, and the administration of the hormones would, it is submitted, almost certainly be regarded as an industrial process.

1 Consumer Protection Act 1987, s 2 (4).

Liability

3.4.4 Actions can be brought against:

(1) the producer, manufacturer or processor of a food and, in certain circumstances, a supplier of ingredients or additives. Packers are not liable unless the packaging alters the essential characteristics of the food;
(2) the importer of food. For the purposes of this Act an importer is one who imports a product into the EEC; not just into the United Kingdom;
(3) where goods are imported into another EC member state and subsequently sold in the United Kingdom, liability rests with the first importer and not the United Kingdom importer;
(4) own-branders. Retailers who put their own name on a product giving the impression that they are the producers are liable. However, words such as 'supplied to . . .' or 'manufactured for . . .' on an own-brand label may avoid such liability subject to the name and address of the manufacturer or supplier being revealed on enquiry by an injured consumer.[1]

1 Consumer Protection Act 1987, s 2 (2).

Defective Products

3.4.5 A product is defective where the safety of that product is not such as persons generally are entitled to expect.[1] When deciding whether a product is defective the courts are required to have regard to all the relevant circumstances including –

* the manner in which the product is marketed;
* any labels, instructions or warnings that are given with it;
* what might reasonably be expected to be done with it;
* the time the producer supplied the product.

In relation to food, these provisions have an obvious and important role in respect of foods for particular nutritional uses. If they fail to fulfil expectations thereby contributing to a nutritional problem there may well be a cause for action.

It has been suggested that errors in nutritional information given voluntarily on labels may be actionable. This is very unlikely for it would be necessary for the plaintiff to prove that injury was caused by the error. This would be difficult to achieve.

Warnings on labels, perhaps associated with minimum durability or use by dates and instructions on storage and use have an important role in minimising liability for potentially hazardous foods, such as cook-chill foods and foods with a very short shelf-life.[1]

1 Consumer Protection Act 1987, s 3.

Time Limit on Actions

3.4.6 A plaintiff must begin his court action within three years of being injured by the defective product. An injured person cannot sue 10 years after the defective product was supplied by the producer.[1]

1 Limitation Act 1980, s 11 as amended by Schedule 1 to the Consumer Protection Act 1987.

Defences

3.4.7 Liability can be avoided if any one of six defences can be proved by the defendant to an action. They are:

(a) the defendant did not supply the product. This is important in cases where counterfeit or out-of-date foods are sold by unauthorised retailers;
(b) the state of scientific and technical knowledge at the time the product was supplied was not such that a producer of products of the same description as the product in question might be expected to have discovered the defect if it had existed in his products while they were under his control. This is the so-called 'development risks' defence and it may be of value in regard to new food analogue products;
(c) the defect was caused by complying with the law. Compliance with a regulation will not necessarily discharge a producer from liability and in order to claim the defence he would have to show that the defect was the inevitable result of compliance.

There has been much debate about whether the use of an additive in a food in conformity with Community law but which caused injury to a consumer could be defended. The answer must surely be in the negative because additives are not required to be used in food;

(d) the defect was not in the product at the time it was supplied. This may be important in cases of bad storage of food by a retailer or where a consumer fails to observe storage instructions on a label or simply keeps the food too long before eating it;

(e) the supplier is not in business. This excludes food made at home by private individuals;

(f) the producer of an ingredient, additive or raw material is not liable if he can show that the defect was due to the recipe for the finished food, or defective specifications being given to him by the producer of the finished food.[1]

1 Consumer Protection Act 1987, s 4.

The Extent of Damages

3.4.8 There is no limit to the damages which may be awarded by a court.

3.5 Some Elementary Precautions for Consumers

3.5.1 Where injury to a person or death or damage to property has resulted from a defective food product procedural mistakes an an early stage can have an adverse effect on any proceedings which may follow. The following points should be borne in mind:

(a) ensure that any remains of the food and its container are retrieved and kept safely;

(b) in cases of injury to a person or death seek immediate medical attendance and, where appropriate, ask the doctor to certify in writing the nature and extent of the injury or the cause of death;

(c) report the matter to the environmental health department of the local food authority (see **2.4**). This is most important in cases where contaminated or defective food may still be on sale. Make it clear to the officer that civil proceedings may follow and that the evidence may be required in court;

(d) if the environmental health officer wishes to take remnants of the food and/or its container away ask him for a written receipt giving details of the food and other evidence taken by him;

(e) unless the injury to a person or damage to property is trivial, and in all cases of death, seek early advice from a solicitor;

(f) whether under the instructions of a solicitor or not, prepare a written statement as soon as possible of the time, date and place of purchase of the food and the events leading up to and immediately following the injury, death or damage;

(g) immediately notify the retailer and the producer (where his identity is known) in writing of the incident, its seriousness, and the intention to seek damages. Where a solicitor has been employed this should be left to him or her;

(h) do not speak to the press, radio or television about the incident. Unjustified allegations or exaggeration by the media attributed to the complainant may prejudice the award or extent of damages.

3.6 Some Elementary Precautions for Producers

3.6.1 In order to minimise the risk of actions for damages there are certain elementary precautions which may be taken by food producers. They are:

(a) ensure that the personnel responsible for receiving and dealing with consumer complaints are properly trained in personal relationships and that they understand the basic principles of criminal and civil liability;
(b) immediately on receipt of a complaint it should be established, preferably in writing, whether or not there has been any death, personal injury or damage to property. The nature and extent of injury or damage should be determined;
(c) the product alleged to be the cause of the problem should be clearly identified by its name, pack size and code number;
(d) the name and address of the retailer who sold the product (assuming that he is not also the producer) should be established;
(e) a full report on the incident should be obtained from the retailer as appropriate;
(f) if the injured person has been medically examined, a copy of the physician's report should be requested;
(g) if possible, samples from the batch from which the product was drawn should be obtained from the retailer for analysis;
(h) if there appears to be any substance to the complaint the matter should be referred to the company's lawyers;
(i) no admissions of any kind nor any offer of compensation should be made until the matter has been considered by lawyers. A chance remark such as 'Oh, we have had several complaints about that', may seriously affect the liability of the company;
(j) if the complaint is of a serious nature and suggests that other products may be defective, product recall procedures should be implemented.

Chapter 4

The Food Safety Act 1990 – Presumptions, Definitions and Interpretation

4.1 The Act

4.1.1 The Food Safety Act 1990 repealed the whole of the Food Act 1984 (save for certain provisions as to markets and fairs and sugar beet). It also repealed the Food and Drugs (Scotland) Act 1956 and amended the Food and Evironment Protection Act 1985. It came substantially into force on 1 January 1991[1] and is now the principal measure for food control in the United Kingdom. It is a criminal measure to be enforced by the ministers and local food authorities.[2] The Act is entirely compatible with the United Kingdom's obligations to the European Economic Community and provides for the implementation and enforcement of EC law.[3]

1 Food Safety Act 1990 (Commencement No 1) Order 1990/1383; Food Safety Act 1990 (Commencement No 2) Order 1990/2372.
2 See Chapter 2.
3 Food Safety Act 1990, s 17.

Application

4.1.2 The Act applies to England, Wales and Scotland. It also applies to the Isles of Scilly subject to such exemptions and modifications as the ministers may by order direct.[1] Her Majesty may, by Order in Council, direct that any of the provisions of the Act shall apply to any of the Channel Islands with such exceptions and modifications as may be specified.[1] Territorial waters of the United Kingdom adjacent to any part of Great Britain are treated as being situated in that part.[2] With certain exceptions the Act does not apply to Northern Ireland but it has been substantially applied to the Province by order.[3] By order the Act can be applied with or without modifications to gas and oil installations as if they were situated in a specified part of the United Kingdom.[4]

1 Food Safety Act 1990, s 57.
2 Food Safety Act 1990, s 58.
3 Food Safety (Northern Ireland) Order 1991/762 (NI 7).
4 Food Safety Act 1990, s 58 (2); Oil and Gas (Enterprise) Act 1982, s 23.

Strict Liability

4.1.3 Most of the offences against the Food Safety Act are of strict liability. It is not necessary for the prosecution to prove guilty knowledge to secure a conviction in relation to the majority of offences. However, revised statutory defences have been provided which are discussed in Chapter 19. See also Chapter 18 for prosecutions.

Regulations and Orders

4.1.4 The Act confers on the ministers substantial powers to make regulations and orders.[1] Conscious of the need to provide a flexible framework for food control, Parliament has empowered the ministers to regulate almost every conceiveable activity in food manufacture, importation, processing, treatment and distribution. It should be borne in mind that the ministers are obliged to consult such organisations as appear to them to be representative of interests likely to be substantially affected by regulations and orders, and such regulations and orders are subject to annulment in pursuance of a resolution of either House of Parliament.[2] It is also pertinent to point out that the bulk of food law made in the United Kingdom is made in implementation of EC requirements. Nevertheless, there is cause for concern about the very extensive enabling powers in the Act and how they may be used if an interventionist-minded government were to be elected.

1 See Chapter 2 and Appendix A.
2 Food Safety Act 1990, s 48 (3) and (4).

4.1.5 Regulations and orders made under the Food Act 1984 or earlier legislation continue in force as if made under the 1990 Act.[1] Substantial modifications have been made to all such subordinate legislation to adapt it to the new provisions of the Act.[2] Regulations made under the European Communities Act 1972 have also been adapted to the provisions of the 1990 Act.

1 Food Safety Act 1990, s 59 (4) and para (2) of Sch 4.
2 Food Safety Act 1990 (Consequential Modifications) (England and Wales) Order 1990/2486; Food Safety Act 1990 (Consequential Modifications) (No 2) (Great Britain) Order 1990/2487.

4.2 Definitions

4.2.1 Definitions in the Food Safety Act 1990 apply to all regulations and orders made under that Act unless expressly excluded.[1] Definitions in regulations and orders apply only to the regulations or order in which they appear, unless the contrary is stated. It is a common misunderstanding in the food industry to assume that definitions in, for example, food labelling regulations apply to regulations concerning particular classes of food and vice versa. They do not, unless, of course, the regulations themselves provide that they shall so apply.

1 Interpretation Act 1978, s 11.

4.3 Exports and Imports

Definitions

4.3.1 In assessing liability under the Food Safety Act 1990 it is important to clearly understand the terms 'export' and 'import'. For the purposes of the Act[1] these terms have the same meaning as in the Customs and Excise Management Act

1979.[2] Thus an 'exporter' in relation to goods for exportation or for use as stores, includes the shipper of the goods and any person performing in relation to an aircraft functions corresponding to those of a shipper. An 'importer', in relation to any goods at any time between their importation and the time when they are delivered out of charge, includes any owner or other person for the time being possessed of or beneficially interested in the goods and, in relation to goods imported by means of a pipe-line includes the owner of the pipe-line.

These definitions therefore apply to goods leaving and entering the United Kingdom. They are not consistent with other legislation on trading matters where the terms apply to goods being brought into or leaving the European Community. In due time, no doubt, these definitions will have to be brought into line with other EC-based legislation.

The definitions are wide and extend liability for certain offences under the Act to agents, shippers and other persons who would not normally expect to be liable for the goods they handle.

1 Food Safety Act 1990, s 53 (1).
2 Customs and Excise Management Act 1979, s 1 (1).

Exemptions for Food for Export

4.3.2 Under most regulations made under or continued in force by the Food Act 1984, food for export from the United Kingdom was exempted. From 26 July 1991 these exemptions were revoked and food for export was required to comply with the regulations unless it could be proved that it complied with the domestic legislation of the importing country.[1] The burden of proof is on the defendant. He must be able to prove that the food does comply with the domestic legislation of the country in which his customer is in business or resides, on the balance of probability. This means, in practical terms, that food exporters should obtain a written statement from an overseas buyer that the food is made to a specification and is labelled in accordance with the relevant legislation.

1 Food Safety (Exports) Regulations 1991/1476.

4.4 Sale, Supply

4.4.1 Most offences against the Food Safety Act 1990 refer to sale or supply, possession for sale, offer or exposure for sale and advertisement for sale. This Act, like its predecessors, extends the meaning of the term 'sale' to include the supply of food, otherwise than on sale, in the course of a business, and any other thing which is done with respect to food and is specified in an order made by the ministers. The Act also applies to food which is offered as a prize or reward or is given away in specified circumstances.[1] The term 'business' is also defined in the Act to include the undertaking of a canteen, club, school, hospital or institution, whether carried on for profit or not, and any undertaking or activity carried on by a public or local authority.[2]

1 Food Safety Act 1990, s 2.
2 Food Safety Act 1990, s 1 (3).

4.4.2 The Act will also bind Crown establishments from 1 April 1992.[1] Thus food supplied in military establishments, HM prisons, public service canteens and the like will be subject to the Act. Note that the Crown will not be criminally liable, but an enforcement authority can apply to the High Court or Court of Session in Scotland for a declaratory judgment that a contravention of the Act has been committed by the Crown. Persons in the public service of the Crown are, however, fully liable.

1 Food Safety Act 1990, s 54 – an order to bring this section into force on 1 April 1992 is imminent.

4.4.3 There is extensive judicial interpretation of the meaning of the term 'sale' in relation to offences against former food law. Much of it is of academic interest only in view of the extended meaning of 'sale' in the 1990 Act. However, it is wise to have regard to the decision that in a self-service store or supermarket there is no contract of sale until payment is made at the check-out.[1]

1 *Pharmaceutical Society of Great Britain v Boots Cash Chemists (Southern) Ltd* [1952] 2 QB 795, [1952] 2 All ER 456.

4.4.4 Exposure for sale means exposure to view in a context where sales are carried out or anticipated. It need not be limited to circumstances where the intending purchaser can actually see it.[1] Offer for sale should not be confused with an offer to treat.[2] The mere display of food does not constitute an offence of 'offering for sale'.

1 *Keating v Horwood* (1926) 90 JP 141; *Wheat v Brown* [1892] 1 QB 418; *McNair v Terroni* [1915] 1 KB 526.
2 *Pharmaceutical Society of Great Britain v Boots Cash Chemists (Southern) Ltd* [1952] 2 QB 795, [1952] 2 All ER 456.

4.4.5 The term 'possession for sale' is particularly important because of new offences created by the 1990 Act which can occur in the factory, packing house or any other premises in addition to retail outlets. Under former food law most criminal offences were linked to sales by retail. By precedent the term 'possession' in the context used in the Act appears to mean possession by the accused or his agent in fact and should be given a popular and not a narrow construction.[1] Physical possession of food by an agent does not divest the owner of possession;[2] but food left for collection by a buyer at a place determined in the contract of sale may not be in the possession of the vendor.[3]

1 *Webb v Baker* [1916] 2 KB 753, 80 JP 449.
2 *Towers & Co Ltd v Gray* [1961] 2 QB 351, [1961] 2 All ER 68; *City Fur Manufacturing Co Ltd v Fureenbond (Brokers) London Ltd* [1937] 1 All ER 799; *R v Sleep* (1861) 25 JP 532, CCR.
3 *Oliver v Goodger* [1944] 2 All ER 481.

4.4.6 Bearing in mind the authorities on the meaning of 'possession' it may reasonably be concluded that food which is in the possession of a food business[1] anywhere in the manufacturing and distribution chain may be in possession for sale. However, it is not clear whether food found in a factory but not yet complete for delivery to a buyer is in possession for sale. In the absence of judicial precedent it is suggested that food found to be defective but in an unfinished state is not in possession for sale.

1 Food Safety Act 1990, s 1 (3).

4.5 Food for Human Consumption

4.5.1 Food is defined in the Act[1] as including drink, articles and substances of no nutritional value which are used for human consumption, chewing gum and other products of a like nature and use and articles and substances used as ingredients in the preparation of food. Specifically excluded from the definition are live animals or birds; live fish which are not used for human consumption while they are alive; fodder or feeding stuffs for animals; birds or fish; controlled drugs; and medicinal products, articles or substances which are licensed under the Medicines Act 1968 unless they are removed from the exemption by an order made by the ministers. Certain enforcement action can be taken on 'food sources' and the term is defined as meaning any growing crop or live animal, bird or fish from which food is intended to be derived (whether by harvesting, slaughtering, milking, collecting eggs or otherwise).

1 Food Safety Act 1990, s 1 (3).

4.5.2 Water is now included in the definition of food where it is used as a drink or ingredient in food thus reversing the exemption in former food statutes. Water delivered by a water undertaker or by private supply for food processing purposes is subject to the Water Industry Act 1991 and regulations made thereunder.[1]

1 Water Supply (Water Quality) Regulations 1989 as amended.

4.5.3 The term 'human consumption' is important because most offences against the Act are concerned with food which has been sold for human consumption or is intended for such sale. 'Human consumption' is defined[1] as including use in the preparation of food for human consumption and thus encompasses food ingredients but, more importantly, there is a presumption[2] that any food commonly used for human consumption when sold or offered, exposed or kept for sale, shall be presumed until the contrary is proved to have been sold or, as the case may be, to have been, or to be intended, for sale for human consumption. Further, any food or article or substance commonly used for human consumption or in the manufacture of food for human consumption which is found on premises used for the preparation, storage or sale of that food or the manufacture of food for human consumption shall, until the contrary is proved, be presumed to be intended for sale or for manufacturing food for sale for human consumption.

1 Food Safety Act 1990, s 53.
2 Food Safety Act 1990, s 3.

4.5.4 These presumptions create a situation where any food sold or found on any premises etc is deemed to be for human consumption unless the person selling or having charge thereof can prove that it is not intended for human consumption. The burden of proof is on the balance of probability[1] but positive evidence is required to rebut the presumption.[2] Thus food on any premises which is not yet in its finished state; or is awaiting acceptance and introduction to the manufacturing process; or has been rejected, should be clearly identified as such so as to contribute to the burden of proof required.

1 *Cant v Harley & Sons Ltd* [1938] 2 All ER 768.
2 *Hooper v Petrou* (1973) 71 LGR 347, [1973] Crim LR 198.

4.6 Food Business/Premises

4.6.1 For the purposes of controlling food businesses and premises certain definitions of and relating to these terms are provided.[1] A 'food business' is defined as meaning any business in the course of which commercial operations with respect to food or food sources are carried out. 'Commercial operations' in relation to any food or contact material covers selling, possessing for sale and offering, exposing or advertising for sale; consigning, delivering or serving by way of sale; preparing for sale or presenting, labelling or wrapping for the purpose of sale; storing or transporting for the purpose of sale; importing and exporting; and, in relation to a food source, means deriving food from it for the purpose of sale or for purposes connected with sale.

1 Food Safety Act 1990, s 1 (3).

4.6.2 'Food premises' means any premises used for the purposes of a food business.'Premises' includes any place, any vehicle, stall or moveable structure and, for such purposes as may be specified in an order made by the ministers, any ship or aircraft of a description so specified.

Chapter 5

Adulteration, Quality and Substitution

5.1 Addition or Abstraction of Substances

5.1.1 Food law owes its origins to the reprehensible practice of adding substances to food to make it go further or the abstraction of substances which may have an alternative use. The addition of water to milk and other drinks, sawdust to bread, sand to pepper and many similar practices were common a hundred years ago. The abstraction of fat from milk, lean from meat and similar practices, although perhaps not as frequent because they involved more effort, were also practised. In more recent times the use of harmful additives in food, residues of medicinal substances administered to animals or pesticides and fungicides in or on animal-based products and crops has posed a danger to the public.

5.1.2 The rendering of food injurious to health by such practices was prohibited in 1875 and that prohibition, only slightly amended, has been re-enacted in the Food Safety Act 1990.[1] It is an offence for any person to render food injurious to health by means of the addition of any article or substance to food, or the use of any article or substance as an ingredient in the preparation of food, or abstracting any constituent from food and subjecting food to any other process or treatment, with intent that it shall be sold for human consumption. In determining whether food is injurious to health a court is required to have regard not only to the probable effect of that food on the health of a person consuming it but also the probable cumulative effect of food of substantially the same composition on the health of a person consuming it in ordinary quantities. The term 'injury to health' includes any impairment, whether permanent or temporary.

1 Food Safety Act 1990, s 7.

5.1.3 To secure a conviction for this offence it is necessary to prove that the food was in fact injurious to health. It has been held that it must be proved that the end product has been rendered injurious to health[1] but the use of an additive which would be harmless to adults but injurious to invalids or children is also an offence.[2] The phrase 'ordinary quantities' limits liability to cases where a person has consumed food in quantities appropriate to his or her sex, size and weight. The probable effects of gluttony would not give rise to proceedings.

1 *Hull v Horsnell* (1904) 68 JP 591.

2 *Cullen v McNair* (1908) 72 JP 280 and *Haigh v Aerated Bread Co Ltd* [1916] 1 KB 878.

5.1.4 The term 'abstract' has given rise to difficulties in the past. It has been held that a failure to correct the natural tendency of a constituent of a fluid to rise to the top or sink to the bottom of a container may amount to abstraction[1] as would the failure to prevent the escape of an evanescent constituent. However, reducing the proportion of a constituent by dilution is not abstraction.[2]

1 *Penrice v Brander* 1921 JC 63; *Bridges v Griffin* [1925] 2 KB 233.
2 *Dearden v Whiteley* (1916) 85 LJKB 1420.

5.1.5 The paucity of judicial interpretation of this offence and the earliness of the dates of revelant cases are because it is difficult to obtain evidence to prove an offence and there are more appropriate and easier offences in other provisions. When the Food Safety Act 1990 was being drafted much consideration was given to whether this offence should be re-enacted. On balance it was decided to do so because, although seldom used, a prohibition on additions or subtractions forms a good foundation for food control.

5.1.6 Examples of cases of cases where this offence may be employed today are:

(a) deliberate contamination (sabotage) of food – although no doubt more dramatic criminal offences will continue to be employed for such offences;
(b) the use of unauthorised additives or additives used far above the maximum levels permitted by regulations – more appropriate offences are available in the Act and in regulations;
(c) leaving residues of pesticide or fungicide in or on food – this is also an offence against regulations and the food safety requirement (see Chapter 10);
(d) the administration of hormones and similar substances to animals – again this is also an offence against regulations.

5.2 Food not of the Nature or Substance or Quality Demanded

5.2.1 It is an offence for a person to sell to the purchaser's prejudice any food which is not of the nature or substance or quality demanded by the purchaser.[1] This offence is basic to the successful control of food and has remained unchanged for many years. It is a remarkably flexible offence and has been used more than any other provision in current or former food law. By virtue of its very wide application this offence has been employed by enforcement authorities in the past to control foreign bodies and mould in food, misdescriptions and a wide variety of other offences, in addition to adulteration and lack of quality. There may be a tendency to continue such practices but there are now more appropriate sanctions for such offences[2] and the ministers have advised enforcement authorities to use them in preference to this offence.[3]

1 Food Safety Act 1990, s 14.
2 Food Safety Act 1990, ss 8 and 15.

3 Statutory Code of Practice No 1: Responsibility for Enforcement of the Food Safety Act 1990, para 18.

Sell

5.2.2 For an offence to arise against s 14 of the Food Safety Act 1990 there must have been a sale for human consumption within the extended meaning of 'sale' – see **4.4.1**. Thus the offence can only be committed on the retail sale of the food concerned and cannot be employed where questions of nature, substance or quality arise from inspection of food on non-retail premises. The need to prove a sale under this section accounts for the practice of authorised officers of enforcement authorities selecting samples for analysis or inspection from the shelves of self-service stores and then taking them to the check-out for payment before beginning the sampling procedures.[1]

1 *Pharmaceutical Society of Great Britain v Boots Cash Chemists (Southern) Ltd* [1952] 2 QB 795, [1952] 2 ALL ER 456.

Prejudice of the Purchaser

5.2.3 For an offence to arise under this section the purchaser must have been prejudiced. This requirement is an expression of the *de minimis* rule (the law is not concerned with trivialities). It is not a defence to such a charge that the purchaser was not prejudiced because he bought the food for analysis or examination.[1]

1 Food Safety Act 1990, s 14.

Meaning of Prejudice

5.2.4 Whether or not a purchaser is prejudiced is a question of fact to be determined in each case. There has been much judicial consideration of this question and under former food statutes provision was made for the display of notices by which the true nature, substance or quality of the food in question could be displayed thus avoiding liability. Such provision was not re-enacted in the 1990 Act but it remains the case that a person cannot be prejudiced if positive information is given as to the true nature, substance or quality of the food. However, such information, no matter how accurate, would not be a defence where compositional standards are linked to the use of certain food names by regulations. Thus, a purchaser would be prejudiced by the sale to him of pork sausages containing only 50% of pork notwithstanding a notice to that effect because regulations[1] require such a product to contain not less than 65% of pork. In such a case there would be an offence of 'quality' under this section and one of deficiency of meat content under the regulations.

1 Meat Products and Spreadable Fish Products Regulations 1984/1566.

5.2.5 Generally speaking, the assessment of prejudice is based on the expectation of the ordinary purchaser. However, it is not necessary to prove actual damage to the consumer. It has been held[1] that the word 'prejudice' cannot be confined to pecuniary prejudice or prejudice arising from the consumption of unwholesome food. The prejudice is that which the ordinary customer suffers, viz, that which is suffered by anyone who pays for one thing and gets another of inferior quality. The words 'to the prejudice of the purchaser' are necessary because if they had not been inserted, a person might have received a superior article to that which he demanded and paid for and yet an offence would have been committed. The price paid for a food is irrelevant when considering whether an offence has been committed[2] although, of course, it might be a factor in considering the seriousness of a case.

1 *Hoyle v Hitchman* (1879) 4 QBD 233.
2 *AJ Mills & Co v Williams* (1964) 62 LGR 354.

5.2.6 In the past the courts have been critical of enforcement authorities who bring proceedings under this section in trivial cases.[1] It should be borne in mind that the 1990 Act now offers a defence of 'all reasonable precautions and all due diligence'[2] which, if properly applied and proved, ought to provide an answer to such cases.

1 *Smedleys Ltd v Breed* [1974] AC 839, [1974] 2 All ER 21, HL; *Sunblest Bakeries Ltd v Andrews* (1985) unreported.
2 Food Safety Act 1990, s 21.

Nature, Substance, Quality

5.2.7 These terms are alternatives and in all proceedings under this section one of them must be chosen. To cite more than one would render an information bad for duplicity.[1]. However, there is some freedom of choice because there is much overlapping between the terms.[2]

1 Magistrates' Courts Rules 1981, r 12; *Bastin v Davies* [1950] 2 KB 579, [1950] 1 All ER 1095.
2 *Preston v Greenclose Ltd* (1975) 139 JP Jo 245.

Nature

5.2.8 This term is usually used where a different kind, species or sort of food is sold from that demanded by the purchaser. Thus cod sold as haddock, giant prunes sold as plums, lumpfish roe sold as caviar and previously frozen meat sold as fresh are examples of 'nature' although other offences against labelling regulations[1] or trade descriptions law[2] would also be approriate. It has been held that reformed white fish sold as scampi[3] could be 'nature' or 'substance' as could minced beef containing small quantities of pork and lamb meat.[4]

1 Food Labelling Regulations 1984/1305, reg 9.
2 Trade Descriptions Act 1968, s 1.
3 *Preston v Greenclose Ltd* (1975) 139 JP Jo 242.
4 *Shearer v Rowe* (1985) 149 JP 698.

Substance

5.2.9 This term is usually taken to mean that the food contains something which is not compatible with the food demanded. It has been widely used to deal with foreign bodies in food and residues of antibiotics in milk etc. Such offences are now more appropriately dealt with under s 8 of the Act (the food safety requirement). However, where a foreign body is sterile and harmless an offence is not necessarily committed even though the foreign body is not of the substance demanded.[1] If the foreign matter is dangerous, such as glass in milk the position must be distinguished from that above.[2]

1 *Edwards v Llaethdy Meirion Ltd* [1957] Crim LR 402.

2 *Southworth v Whitewell Dairies Ltd* (1958) 122 JP 322.

Quality

5.2.10 This means 'commercial quality' and not merely description.[1] Failure of a food to satisfy statutory compositional standards is an offence under this provision and where there is no statutory standard a court may be invited to set a standard based on what it believes an ordinary purchaser would expect to receive. Evidence of a public analyst to the effect that a food is not of the quality demanded is valuable as a guide to proceedings but it is not conclusive. The test is what the purchaser expected to receive and it is a question of fact not of opinion.[2] However, if the public analyst's opinion is undisputed and is to the effect that the food is not of the quality demanded then there should be a conviction.[3]

1 *Anness v Grivell* [1915] 3 KB 685.
2 *Goldup v John Manson Ltd* [1982] QB 161, [1981] 3 All ER 257; *McDonald's Hamburgers Ltd v Windle* (1986) 151 JP 333; *Collins Arden Products Ltd v Barking Corpn* [1943] KB 419, [1943] 2 All ER 249.
3 *Williams v Hurrell's Stores Ltd* [1954] Crim LR 543; *Bowker v Woodroffe* [1928] 1 KB 217.

The Food Demanded

5.2.11 Where there is a statutory standard the purchaser is deemed to have demanded a food to that standard.[1] In cases where a breach of a statutory standard occurs it is usual to bring proceedings under the regulations setting the standard and, if appropriate, under s 14 of the Act also.

1 The logic of this proposition is that if a buyer demands a food for which there is a statutory standard, but receives an inferior product, he is prejudiced and the offence is committed.

5.2.12 Where there is no statutory standard the court must judge what it was that the purchaser in fact demanded based on the evidence to hand. In two leading cases, both concerning the quantity of fat in minced meat,[1] the Divisional Court held that the question as to what was demanded should be answered after consideration or what the purchaser actually said, any information or disclaimers as to quality by the seller, whether there was prejudice and whether different qualities of the food demanded were available at different prices.

1 *Goldup v John Manson Ltd* [1982] QB 161, [1981] 3 All ER 257; *TW Lawrence & Sons Ltd v Burleigh* (1981) 146 JP 134; – see also *Morton v Green* (1881) 8 R 36; *Anness v Grivell* [1915] 3 KB 685; *Hunt v Richardson* [1916] 2 KB 446; *Collins Arden Products Ltd v Barking Corpn* [1943] KB 419, [1943] 2 All ER 249; *Bowker v Woodroffe* [1928] 1 KB 217, [1927] All ER Rep 415.

5.3 Compositional Standards

5.3.1 Legislation requiring a minimum content of certain ingredients in basic foods or prohibiting the use of certain undesirable ingredients can be traced back to the early 19th century when governments, fearful of declining dietary standards for the population, thought it wise to impose such requirements. Modern food standards owe their origins to food shortages during the Second World War and it is only in very recent times that some of the war-time standards have been revoked or revised. Until the early 1970s most food standards were mandatory, that is foods conforming to a definition in regulations were obliged to be made to a minimum standard laid down no matter what description was applied to them. In more recent times the concept of reserved descriptions has been devised under which a food is obliged to be made to prescribed standards if certain reserved descriptions are used in the name of the product. The distinction between compulsory standards and those linked to reserved descriptions is largely academic because in practice reserved descriptions are the only names under which foods subject to them can be sold to satisfy the requirements of labelling regulations that a food name must indicate the true nature of the food (see Chapter 6). The distinction is also confusing because certain regulations refer to reserved descriptions but then make their use compulsory.

Recently, the term 'restricted name' has been introduced into regulations. This means that a name so restricted can only be used when the food satisfies a definition laid down for it but it does not impose any compositional standards.

5.3.2 *Foods for which compositional standards* have been introduced (in alphabetical order) are:

Bread – The Bread and Flour Regulations 1984/1304, as amended – UK domestic law.

The words 'wholemeal', 'brown', 'wheat germ', 'white', 'aerated' and 'part-baked' or 'partly baked' must be used in the name of the product where the product falls within the definition of bread. Use of a name including these words for a product not satisfying certain compositional standards is unlawful. Use of the name 'wheatmeal' to describe any bread is prohibited.

Certain ingredients which need not be included in the product name are laid down, as are ingredients whose use is compulsory and additives which may be used up to certain prescribed maxima or, where there is no prescribed maximum, to levels consistent with good manufacturing practice.

Butter – The Butter Regulations 1966/1074 as amended – UK domestic law to be replaced when Community rules on yellow fats are adopted.

It is unlawful to use the word 'butter' to describe any product unless it conforms to the definition of butter and satisfies composition standards as to the milk fat, milk solids other than fat, and water content.

The use of the word 'butter' for products consisting wholly or partly of non-dairy ingredients has long been controversial and the subject of litigation. The matter has been put beyond doubt now by Council Regulation (EEC) No 1898/87 given the force of law in the United Kingdom by the Milk and Milk Products (Protection of Designations) Regulations 1990/607.

Caseins and caseinates – The Caseins and Caseinates Regulations 1985/2026 in implementation of Council Directive 83/417/EEC.

The words 'edible acid casein', 'edible rennet casein' and 'edible caseinates' are reserved descriptions. It is unlawful to use them for any product unless it complies with prescribed heat treatment processes and technological adjuvants and bacterial cultures.

Cheese – The Cheese Regulations 1970/94, as amended – UK domestic law long overdue for revision but likely to remain in force unchanged until Community measures on cheese are adopted.

Cheddar, Blue Stilton, Derby, Leicester, Cheshire, Dunlop, Gloucester, Double Gloucester, Caerphilly, Wensleydale, White Stilton, Lancashire, Edam, Loaf Edam, Baby Edam, Baby Loaf Edam, Gouda, Baby Gouda, Danablu, Danbo, Havarti, Samsoe, Emmental or Emmentaler, Gruyere or Greyerzer or Gruviera, Tilsiter or Tilsit or Tylzychi, Limburger, Saint Paulin, Svecia and Provolone are reserved descriptions for which minimum percentages of milk fat in dry matter and maximum percentages of water calculated on the total weight of the cheese are laid down. For other cheeses the reserved descriptions 'full fat hard cheese', 'medium fat hard cheese', 'skimmed milk hard cheese', 'full fat soft cheese','medium fat soft cheese', 'low fat soft cheese', 'skimmed milk soft cheese', 'full fat whey cheese', 'whey cheese', 'skimmed whey cheese', 'full fat processed cheese', 'medium fat processed cheese', 'skimmed milk processed cheese' and 'cheese spread' or 'cheese food' must be used and minimum milk fat and maximum water contents are prescribed.

Permitted ingredients are laid down and certain labelling requirements must be satisfied.

Chocolate Products – The Cocoa and Chocolate Products Regulations 1976/541, as amended – implementation of Council Directives 74/411/EEC, 74/644/EEC and 75/155/EEC – perhaps the most badly drafted and incomprehensible regulations currently in force.

Reserve descriptions with quality standards for cocoa solids, cocoa butter and milk solids are laid down for chocolate, plain chocolate, chocolate vermicelli, chocolate flakes, gianduja nut chocolate, couverture chocolate, milk chocolate, milk chocolate vermicelli, milk chocolate flakes, gianduja nut milk chocolate, couverture milk chocolate, white chocolate, filled chocolate, a chocolate or chocolates, cream chocolate and skimmed milk chocolate. Permitted additional ingredients are listed and there are labelling requirements as to the cocoa and milk solids content.

Cocoa Products – The Cocoa and Chocolate Products Regulations 1976/541 – as above.

Reserve descriptions with quality standards for maximum water, minimum cocoa butter, cocoa butter mixed with sucrose and fat reduced cocoa when mixed with sucrose are laid down for cocoa, cocoa powder, fat reduced cocoa, fat reduced cocoa powder, sweetened cocoa, sweetened cocoa powder, drinking chocolate, sweetened fat reduced cocoa, sweetened fat reduced cocoa powder and fat reduced drinking chocolate. There are also standards for bean, dust, nib, fines, mass, press cake, fat-reduced press cake, expeller press cake, butter, press butter, expeller butter, refined butter and fat.

Coffee Products – Coffee and Coffee Products Regulations 1978/1420, as amended – implementation of Council Directive 77/436/EEC.

Compulsory reserve descriptions for coffee, coffee mixtures, coffee extracts, chicory extracts and extracts of blends of coffee and chicory are laid down together with definitions thereof and certain minima and maxima coffee and chicory dry matter contents. There are compositional requirements for raw materials, specified additional ingredients and labelling declarations as to decaffeinated products and certain manufacturing processes.

Cream – The Cream Regulations 1970/752, as amended – UK domestic law – likely to remain in force until EC requirements are adopted.

Compulsory descriptions 'clotted cream', 'double cream', 'whipping cream', 'whipped cream', 'sterilised cream', 'cream' or 'single cream', 'sterilised half-cream' and 'half cream' are laid down with associated minimum milk fat content. Permitted ingredients are also prescribed.

Fish cakes – The Food Standards (Fish Cakes) Order 1950/589 – UK domestic law – soon to be revoked when new comprehensive fish products regulations are introduced.

The Order prescribed a minimum fish content of 35% but the new regulations are unlikely to re-enact this requirement.

Flour – The Bread and Flour Regulations 1984/1304 – see under 'bread' above.

The names under which flour may be sold, including 'wholemeal', 'brown flour', 'flour' and 'self-raising flour' with appropriate definitions and/or compositional requirements are laid down. Subject to certain exemptions wheat flour must be fortified with specified nutrients and the use of certain additional ingredients and additives is controlled.

Fruit Juices and Fruit Nectars – The Fruit Juices and Fruit Nectars Regulations 1977/927, as amended – implement Council Directive No 75/726/EEC.

Definitions and compositional standards as to fruit content and acidity are prescribed for fruit juice, concentrated fruit juice, dried fruit juice and fruit nectar. The names used must be qualified by an indication of the type of fruit from which the food is prepared. Labelling must include certain indications of fruit content, sweetening and concentration.

Honey – The Honey Regulations 1976/1832, as amended, – implement Council Directive 74/409/EEC.

Use of the word 'honey' is restricted to substances which satisfy the definition of 'honey' and satisfy compositional requirements as to apparent reducing sugar content, moisture content, apparent sucrose content, water-insoluble solids content and ash content. Additional labelling requirements are imposed and references to the blossom or plant origin of the honey or regional, topographical or territorial origin may only be applied when the honey is wholly or mainly produced as indicated.

Ice-cream – The Ice-Cream Regulations 1967/1866, as amended – UK domestic law – likely to remain in force until EC requirements adopted.

The regulations specify compositional requirements as to milk fat and milk solids non-fat for ice-cream and Parev Ice and specify labelling requirements for them.

Jam and Similar Products – Jam and Similar Products Regulations 1981/1063, as amended, – implement Council Directive 79/693/EEC.

Reserved descriptions with associated quality standards as to the quantity of fruit pulp and soluble solids in the product are applied to extra jam, jam, extra jelly, jelly, marmalade, sweetened chestnut puree, reduced sugar jam, reduced sugar jelly, reduced sugar marmalade, UK standard jelly, 'X' curd, 'A' lemon cheese, 'Y' flavour curd and mincemeat. Sweetening agents and the sulphur dioxide content are controlled; permitted additional ingredients are prescribed and additional labelling requirements are imposed.

Margarine – Margarine Regulations 1967/1867, as amended – UK domestic law – likely to be replaced in due course by EC rules on edible fat spreads.

Compulsory fat and vitamin contents are laid down for any product sold as 'margarine'. The labelling of mixtures of dairy ingredients with margarine is controlled. Attempts to compare margarine products favourably with butter have often been controversial and the subject of litigation.[1] The matter should now be considered in the light of Council Regulations (EEC) No 1898/87 and the Milk and Milk Products (Protection of Designations) Regulations 1990/607.

Meat Products and Spreadable Fish Products – the Meat Products and Spreadable Fish Products Regulations 1984/1566, as amended – United Kingdom domestic law – EC rules as to meat products are pending.

The regulations prohibit the use of reserved descriptions unless requirements as to meat or fish contents are satisfied. The products covered are burgers, economy burgers, hamburgers, chopped meats, corned meats, luncheon meats, meat pies and meat puddings, Melton Mowbray pie, game pie, Scottish pie or Scotch pie, pies or puddings, pastys or pasties, bridies, sausage rolls, sausages, links, chipolatas and sausage meat, pastes, pates and spreads. It is necessary to declare meat and/or added water contents and there are requirements as to the ratio of lean and fat meat. The use of offals is controlled.

Milk – The Milk and Dairies (Channel Islands and South Devon Milk) Regulations 1956/919; the Drinking Milk Regulations 1976/1883 giving effect to Council Regulation (EEC) No 1411/71; Milk and Dairies (Milk Bottle Caps) (Colour) Regulations 1976/2186.

The regulations prohibit the sale of drinking milk other than raw milk, non-standardised whole milk produced in the UK, standardised whole milk imported from another EC country having a fat content figure not less than a guideline figure, semi-skimmed milk and skimmed milk. For standardised whole milk the guideline figure is fixed annually as the weighted average fat content of the whole milk produced and marketed in the UK during the previous milk year. For all other grades of milk, minimum fat contents are laid down in the regulations.

The cap colour for bottles containing pasteurised, sterilised, UHT and untreated milk are also laid down.

Milk-based Drinks – The Milk-based Drinks (Hygiene and Heat Treatment) Regulations 1983/1508, as amended – UK domestic law but made pursuant to the European Court of Justice judgment in *Re Ultra Heat Treated Milk; EC Commission v United Kingdom* [1983] 2 CMLR 1.

The regulations require that a liquid drink (other than a fermented drink) comprising a minimum of 85% milk and other permitted ingredients shall be made from cows' milk (whether or not separated) but not cream separated from milk, dried milk, condensed milk, evaporated milk or butter milk. Permitted ingredients in such a product are listed and certain processing standards apply.

Milk – Condensed and Dried – The Condensed Milk and Dried Milk Regulations 1977/928, as amended – implement Council Directive No 76/118/EEC.

The regulations prescribe definitions and reserved descriptions for condensed milk and dried milk products with compositional standards for milk fat, non-milk solids and, for dried milk products, the maximum moisture content. Permitted ingredients are prescribed and special labelling requirements are laid down.

Natural Mineral Waters – The Natural Mineral Waters Regulations 1985/71, as amended – implement Council Directive No 80/777/EEC.

The regulations require all natural mineral waters to be officially recognised; prescribe conditions for the exploitation of natural mineral water sources; regulate the treatment of natural mineral waters; lay down microbiological criteria for them; prohibit the presence of toxic substances; lay down commercial designations; provide for safe packaging; and impose labelling requirements.

Poultry – The Poultry Meat (Water Content) Regulations 1984/1145, as amended – implement Council/Commission Regulations (EEC) 1691/77, 361/78, 2238/78, 641/79, 2632/80, 3134/81 and 3204/83.

The regulations lay down the maximum water content of frozen poultry carcases.

Scotch Whisky – The Scotch Whisky Order 1990/998, made under the Scotch Whisky Act 1988 – UK domestic law.

The Act provides for injunctive action where a product is described as 'scotch whisky' but fails to satisfy the definition and prescribed alcoholic strength of Scotch whisky contained in the Order.

Soft Drinks – The Soft Drinks Regulations 1964/760, as amended – UK domestic legislation.

The regulations prescribe reserved descriptions linked to compositional requirements as to fruit and sugar content with special labelling requirements. New regulations will be made very soon which will require declarations of fruit or fruit juice content. Restricted names for squash, crush, Indian tonic water, quinine tonic water and soda water linked to definitions thereof will be introduced but there will be no compositional standards save for restrictions on certain claims and control of additives.

Spirit Drinks – Council Regulation (EEC) No 1576/89 given effect by the Spirits Drinks Regulations 1990/1179 – EC Marketing Standards Legislation.

Definitions with minimum alcoholic strength for rum, whisky/whiskey, grain spirit, wine spirit, brandy, grape marc spirit, fruit spirit, cider/perry spirit, gentian spirit, fruit spirit drinks, juniper-flavoured spirit drinks, caraway-flavoured spirit drinks, aniseed-flavoured spirit drinks and liqueur and spirit drinks are prescribed.

Sugar Products – The Specified Sugar Products Regulations 1976/509, as amended – implement Council Directive No 73/437/EEC.

The regulations prescribe definitions, methods of analysis and reserved descriptions linked to dry matter, D-glucose, dextrose and sulphated ash contents. Special labelling requirements are imposed with restrictions on the use of the word 'white'.

1 *Van den Berghs and Jurgens Ltd v Independent Broadcasting Authority* (1976) unreported.

EC Policy on Compositional Standards

5.3.3 In November 1985 the EC Commission issued a communication[1] in which it outlined its policy on food law up to the completion of the internal market in 1992. Included in this document was a declaration that henceforward the policy of the EC would be away from compositional standards ('recipe law' as it is called in Brussels) and towards a more informative system of food labelling. This policy was largely dictated by the impossibility of obtaining agreement among the member states as to what standards should be applied by way of recipe law. It was therefore declared that no further compositional standards would be made and those currently existing would in due course be swept away. In their place a system of quantitative indications of ingredients (QUID) would be introduced when the labelling directive[2] was revised. In August 1989 the United Kingdom government issued a memorandum supporting this policy. Since 1985 no new compositional standards have been adopted by the EC and attempts by some member states to introduce them into national legislation have been blocked by the Commission.

In November 1991 the first working papers on QUID were issued containing proposals that the major two or three ingredients in a food, as appropriate and any characterising ingredients should be quantitatively indicated.

1 COM (85) 603 Final.
2 Council Directive 79/112/EEC (OJ No L 033; 08.02.79) – *Butterworths Law of Food and Drugs.*

5.3.4 *The future of existing compositional standards* is the subject of considerable controversy in Brussels and in representative organisations in the United Kingdom. Those who wish to see them maintained believe that their abolition would result in a decrease of quality in the foods concerned. They point out that consumers have become accustomed to the standards and would be disadvantaged if manufacturers were free to produce such foods to a lower quality no matter what additional information may be provided on labels. Certain manufacturers also support the retention of such standards because they believe that they result in fairer competition between them. Others accept that the maintenance of national standards is impossible in the single market and believe that QUID will provide adequate protection for consumers against reductions in quality.

Although United Kingdom compositional standards can be enforced in respect of foods manufactured and sold in the United Kingdom they cannot be enforced with respect to foods imported from another member state. The principle that a product which may be lawfully marketed in the exporting member state may not be refused entry into any other Member State established in the *Cassis de Dijon* judgment of the European Court (see **1.4.1**) and now enshrined in the Single European Act means that competing products imported into the United Kingdom are not obliged to be made to the same standard as the home product. They must, of course, be labelled to distinguish them from the domestic product (see Chapter 6). It remains an open question as to whether a prosecution for breach of a compositional standard based on United Kingdom domestic law could be defended on the grounds that it is contrary to the EC Treaty.

It is now near certain that all compositional standards will be revoked, whether derived from EC or national legislation, by the end of 1993. The proposed directive on QUID is under active consideration and is likely to require that the major one or two ingredients in a wide range of foods will have to be quantitatively indicated on labels as will certain characterising ingredients. Food manufacturers in the United Kingdom already have more experience of QUID than their competitors in other EC member states by virtue of meat, fruit juice, cocoa solids and a variety of other content declaration requirements in existing law.

Marketing Standards

5.3.5 Notwithstanding the policy indications discussed above, the EC Commission would appear to have no inhibitions about creating new compositional standards in defence of particular food sectors. Marketing standards for horticultural produce, wine and milk have been in force almost since the foundation of the EC but in recent times new ones have been imposed by regulations for oils and fats, poultrymeat, spirit drinks, dairy products, fish products and alcoholic beverages. Marketing standards for horticultural produce and fish products, for example, impose requirements as to the size, shape and appearance of the food but others, such as milk and spirit drinks, impose compositional requirements which are indistinguishable from compositional standards for consumer protection purposes. It is an offence not to comply with such standards and the Commission apparently sees no contradiction in a policy which calls for the abolition of standards for consumer protection but their introduction for trade protection.

Chapter 6

Labelling, Presentation and Advertising

6.1 False or Misleading Labels and Advertisements

6.1.1 It is an offence to give with any food sold, or to display with any food offered or exposed for sale, or in possession for sale, a label, whether or not attached to or printed on the wrapper or container, which falsely describes the food or which is likely to mislead as to the nature or substance or quality of the food.[1]

1 Food Safety Act 1990, s 15 (1).

6.1.2 It is also an offence to publish, or to be a party to the publication of, an advertisement which falsely describes any food or which is likely to mislead as to the nature or substance or quality of any food.[1]

1 Food Safety Act 1990, s 15 (2).

A Label or Advertisement

6.1.3 These offences are basic to the prevention of false or misleading labels or advertisements and are very wide in their application. The word 'label' is not defined but by a definition thereof in the Food Labelling Regulations 1984[1] 'labelling' includes any words, particulars, trade mark, brand name, pictorial matter or symbol relating to the food and appearing on the packaging of the food or in any document, notice, label, ring or collar accompanying food. That definition is not, of course, directly binding as to this offence but the use of the words 'whether or not attached to or printed on the wrapper or container' suggest that any information given visually by words or illustrations could give rise to an offence. A label would not include a verbal statement but the definition of 'advertisement'[2] includes any notice, circular, label, wrapper, invoice or other document, and any public announcement made orally or by any means of producing or transmitting light or sound.

It should also be borne in mind that a verbal statement can be a false trade description under trade descriptions law[3] and the Trade Descriptions Act 1968 is often employed by enforcement authorities instead of or in addition to proceedings under the Food Safety Act 1990 where appropriate.

For all practical purposes it should be assumed that any false or misleading statement as to food for human consumption, however given, is an offence.

1 Food Labelling Regulations 1984/1305, reg 2 (1).
2 Food Safety Act 1990, s 53 (1).
3 Trade Descriptions Act 1968, ss 1 and 4 (2).

Falsely Describes the Food

6.1.4 Prosecuting authorities have not employed this offence as frequently as the alternative 'likely to mislead . . .' because the term 'false' is stronger and more difficult to prove. It is necessary to prove that a label or advertisement is explicitly false, that is, wholly untrue. However, a court is not precluded from finding that a label or advertisement is false or misleading even if it contains an accurate statement of the composition of the food.[1] The concept of falseness on food labels has been considered by the appellate courts on numerous occasions but nearly always with regard to civil litigation or offences against the trade descriptions law. Bearing in mind that the term 'a false trade description' in the Trade Descriptions Act 1968 includes a misleading statement[2] and the burden of proof of falseness in civil litigation is different from that in criminal law, the precedents indicated below should be treated with caution:

(a) use of the word 'port' for a product not from Portugal or the island of Madeira has been held to be a false trade description;[3]
(b) it is a false description to label a mixture of English and Tarragona wine as 'Fine British Tarragona Wine' notwithstanding the fact that it would mislead no one;[4]
(c) a solution of acetic acid and caramel is falsely described if labelled as 'vinegar';[5]
(d) the use of the name 'sherry' for a product not from the Jerez de la Frontera area of Spain is a false description;[6]
(e) a fruit juice described as natural was not falsely described even though it had been reconstituted in part from concentrated juice and had been pasteurised.[7]

1 Food Safety Act 1990, s 15 (4).
2 Trade Descriptions Act 1968, s 3 (2).
3 *Sandeman v Gold* [1924] 1 KB 107.
4 *Holmes v Pipers Ltd* [1914] 1 KB 57.
5 *Kat v Diment* [1951] 1 KB 34, [1950] 2 All ER 657.
6 *Vine Products Ltd v Mackenzie & Co Ltd* [1969] RPC 1.
7 *Amos v Britvic Ltd* (1984) 149 JP 13.

Likely to Mislead

6.1.5 It has been held that a label may be false on account of what it omits even thought it may be factually true.[1] In matters as complex and variable as food labelling these precedents may be unsafe and for all practical purposes it is suggested that the offence of 'falseness' should be used only where there is a misstatement of fact. All other offences are more appropriate to a charge of 'likely to mislead'.

As mentioned above, a label or advertisement may be misleading even though it contains a factually correct statement of the composition of the food. Thus a label which bears an accurate list of ingredients in compliance with labelling requirements[2] may be misleading by virtue of the product name, instructions for use, processes or treatments or background illustrations. Thus a label which stated 'fully prepared, sliced, selected tinned apples (unsweetened)' was held to be misleading because the fruit had lost about 25% of its original solids because of added water.[3]

This offence has been employed, generally unsucessfully, to prevent the alleged misuse of traditional food names and terms. A product labelled '2 Chicken Breast Steaks – Flaked and formed chicken in a crispy crumb' was held on appeal to the Crown Court not be to be misleading. The enforcement authority had attempted to restrict the use of the word 'steak' to a solid piece of muscle meat.[4] Similarly, a product labelled 'Pura Vegetable Lard' with a subsidiary statement '100% vegetable oils' was held not to be misleading despite the fact that lard is a product wholly of pig fat.[5] On the other hand, the labels of products named 'Elmlea Single' and 'Elmlea Whipping' both bearing the additional words 'the real alternative to cream' but which were made from vegetable oil and packed in cartons resembling those usually used for cream were held on appeal to the crown court to be misleading under the Trade Descriptions Act 1968.[6] The information given on the labels was factually correct but the impression created for the ordinary consumer may have been that the product consisted of cream.

1 *R v Lord Kylsant* [1932] 1 KB 442; *R v Bishirgian* [1936] 1 All ER 586; *Curtis v Chemical Cleaning and Dyeing Co Ltd* [1951] 1 KB 805, [1951] 1 All ER 631, CA.
2 Food Labelling Regulations 1984, regs 6(b) and 13–19.
3 *Arlidge v Blue Cap Foods (Kent) Ltd* (1965) 63 LGR 167.
4 *GW Padley (Poultry) Ltd v Elkington* (1986) unreported.
5 *Wolkind and Northcott v Pura Foods Ltd* (1987) 151 JP 492.
6 *Burleigh v Van Den Berghs & Jurgens Ltd* [1987] BTLC 337.

6.2 Presentation

6.2.1 It is unlawful to sell, offer or expose for sale, or have in possession for the purpose of sale, any food the presentation of which is likely to mislead as to the nature or substance or quality of the food.[1] This rather curious requirement of the Labelling Directive[2] was originally enacted and remains an offence under food labelling regulations[3] but the opportunity was taken in drafting the Food Safety Act 1990 to include it in primary legislation. It should be noted that the offence in the labelling regulations is mitigated by a *de minimis* provision that the presentation must be misleading to a 'material degree'. This provision does not appear in the Act.

1 Food Safety Act 1990, s 15 (3).
2 Council Directive 79/112/EEC as amended.
3 Food Labelling Regulations 1984, reg 4.

Definition of Presentation

6.2.2 Presentation is defined[1] as including the shape, appearance and packaging of the food, the way in which the food is arranged when it is exposed for sale and the setting in which the food is displayed with a view to sale, but does not include any form of labelling or advertising. It therefore follows that truthful labelling can mitigate or negate any misleading presentation.

1 Food Safety Act 1990, s 53 (1).

Problems of Presentation

6.2.3 There have been a few prosecutions for misleading presentation but the offence is not common. Cases have included fatty mince displayed under red lighting to give a misleading impression as to its fat content; meat containing novel protein displayed together with whole meat; analogue dairy products being displayed with genuine milk-based products; and artificially-flavoured fruit products being packed in fruit-shaped containers.

6.3 Advertising

6.3.1 In addition to the offence as to the advertising of food discussed above, non-broadcasting advertising is subject to the self-regulation system of the advertising industry. The British Code of Advertising Practice[1] contains specific restrictions as to claims about aids to dieting, health claims, low-calorie claims and references to vitamins. The Code is administered by the Advertising Standards Authority from whom copies of the code can be obtained.[2]

1 *British Code of Advertising Practice* (8th Edn) December 1988.
2 Advertising Standards Authority, Brook House, 2–16, Torrington Place, London, WC1E 7HN; Tel: 071–580–5555.

6.3.2 Food advertisements which are broadcast by television are subject to the advertising rules laid down by ITC Code of Advertising Standards and Practice.[1] The Independent Television Commission is under a statutory duty[2] to draw up and enforce a code governing standards and practice in television advertising. Enforcement is by way of pre-vetting advertisements before transmission which is currently the responsibility of the Independent Television Association.[3] Guidance notes on certain aspects of food advertising are available from the ITV Association.

1 ITC Code of Advertising Standards and Practice; Advertising and Sponsorship Division, ITC, 70 Brompton Road, London, SW3 1EY; Tel: 071–584–7011.
2 The Broadcasting Act 1990.
3 ITV Association, Knighton House, 56 Mortimer Street, London W1N 8AN.

6.4 Food Labelling

6.4.1 Food labelling throughout the 12 member states of the EC is controlled by the Labelling Directive.[1] The Directive has been implemented in the United Kingdom by the Food Labelling Regulations 1984, as amended.[2] Although primarily intended as an instrument of consumer protection, the Directive has an important role in ensuring fair competition between food manufacturers in the member states. By virtue of the mutual recognition of products enshrined in the Treaty of Rome and given expression in the *Cassis de Dijon* judgment of the European Court of Justice[3] it is essential that food imported from other member states should be properly labelled to distinguish it from competing products in the importing country. It is therefore unfortunate that despite the fact that the Directive has been in operation in the Community for more than 10 years (Portugal, Spain and Greece excepted) labelling standards on many food products imported into the United Kingdom are unsatisfactory.

Although food labelling is the subject of much criticism by consumer bodies, compliance with the Directive in the United Kingdom is high. This has much to do with the diligence of British food manufacturers and the excellent work of enforcement officers over the past 10 years.

1 Council Directive 79/112/EEC as amended.
2 SI 1984 no 1305 – for amendments see Appendix A.
3 See **1.4**.

6.4.2 In 1990 the ministers asked the Food Advisory Committee (FAC) to review the law on food labelling and the committee's report[1] recommended a number of important changes. The government has indicated its support for most of the recommendations but because any change in the law must be agreed by the EC, implementation of the recommendations may be long delayed. However, where appropriate the proposals are mentioned in the following summary of food labelling law.

1 Food Advisory Committee Report on its Review of Food Labelling and Advertising 1990 (HMSO).

The Regulations

6.4.3 The Food Labelling Regulations 1984 apply to all food for human consumption save for milk, which is controlled by Council Regulation (EEC) No 1411/71 and United Kingdom regulations applicable thereto (see Appendix A), and natural mineral waters (not artificially carbonated) which are subject to Council Directive 80/777/EEC and the Natural Mineral Waters Regulations 1985. The detailed labelling provisions of the regulations do not apply to:

(a) specified sugar products;
(b) cocoa and chocolate products;
(c) honey;
(d) condensed milk products and dried milk products;
(e) hen eggs;
(f) wine and grape musts;
(g) additives labelled and sold as such;
(h) food prepared on domestic premises for sale for the benefit of the person preparing it by a society registered under the Industrial and Provident Societies Act 1965;
(i) food prepared otherwise than in the course of a trade carried on by the person preparing it;
(j) coffee, coffee mixture, coffee extract product, chicory extract product or other such designated product which is ready for delivery to a catering establishment.

The labelling of the above products is controlled by other regulations for which see Appendix A.

General Labelling Requirements

6.4.4 The general labelling requirement laid down by Part III of the Food Labelling Regulations 1984 for non-exempted foods is that the food shall be marked or labelled with[1] –

(a) the name of the food;
(b) a list of ingredients;
(c) the appropriate durability indication;
(d) any special storage conditions or conditions of use;
(e) the name or business name and an address or registered office of the manufacturer or packer, or of a seller established within the EEC;
(f) particulars of the place of origin of the food if failure to give such particulars might mislead a purchaser to a material degree as to the true origin of the food; and

(g) instructions for use if it would be difficult to make appropriate use of the food in the absence of such instructions.

1 Food Labelling Regulations 1984, reg 6.

Food Product Names

6.4.5 Subject to a few exemptions, all food must be identified by a name whether it is prepacked or not. There are three classes of name which may be lawfully used. They are (i) a name prescribed by law, or (ii) a customary name, or (iii) a name which indicates the true nature of the food:

(i) *Names prescribed by law*[1] – where there is a name prescribed by law that name must be used on the label of the food or, if the food is not prepacked, on a ticket or notice Such a name may be qualified by other words which make it more precise unless the regulations which prescribe the name state otherwise. Foods for which names are prescribed by law are:

fish, melons (the species must be given), potatoes (the variety must be given) and vitamins;[2]
soft drinks – the Soft Drinks Regulations 1964;
cheese – the Cheese Regulations 1970;
cream – the Cream Regulations 1970;
honey – the Honey Regulations 1976;
fruit juices and fruit nectars – the Fruit Juices and Fruit Nectars Regulations 1977;
coffee and coffee products – the Coffee and Coffee Products Regulations 1978;
bread and flour – the Bread and Flour Regulations 1984;
natural mineral waters – the Natural Mineral Waters Regulations 1985.

For references as to the above regulations see Appendix A.

For certain foods reserved descriptions are laid down by regulations. They are not names required by law but names which, if used, create a requirement to meet compositional standards. For example, a 'salmon spread' would have to contain not less than 70% of fish because the word 'spread' is a reserved description. A product named 'salmon mousse' would not have to satisfy the standard because the word 'mousse' is not a reserved description.

The products for which reserved descriptions are prescribed are:

butter – the Butter Regulations 1966;
ice cream – the Ice Cream Regulations 1967;
margarine – the Margarine Regulations 1967;
cocoa and chocolate products – the Cocoa and Chocolate Products Regulations 1976;
condensed and dried milk – the Condensed and Dried Milk Regulations 1977;
jam etc – the Jam and Similar Products Regulations 1981;
meat products and spreadable fish products – the Meat Products and Spreadable Fish Products Regulations 1984.

For a discussion of the future of compositional standards and the use of reserved descriptions see **5.3**. In practice it is difficult to devise a different name for a product with a reserved description to satisfy the requirements as to 'true nature' etc discussed below.

(ii) *Customary Names*[3] – where there is no name prescribed by law a customary name may be used. A customary name is one which is customary in the area where the food is sold. Such names may include 'Cornish Pasty', 'Welsh Rarebit', 'Brighton Rock', 'Lancashire Hot Pot' and so on. Many foreign food names have become common in the United Kingdom over the years and may qualify as customary names. Examples are 'muesli', 'lasagne', 'macaroni' and 'petit fours'.

Whether a food name is customary in a given area is a question of fact. Some food manufacturers are tempted to apply geographical words to a common product or to search out old-fashioned and seldom used names which have a marketing appeal in the belief that they will qualify as customary names. In such cases it is wise to add an additional name satisfying the requirements as to indication of true nature discussed below.

The Food Advisory Committee has recommended that the option to use a customary name should be abolished. Note also that a trade mark, brand name or fancy name may not be used as a substitute for the name of the food.[4]

(iii) *Indication of the true nature of the food* [5] – If there is no name prescribed by law and if there is no customary name or if the customary name is not used, the name of the food must indicate the true nature of the food and must be sufficiently specific to enable the food to be distinguished from other food with which it could be confused and, if necessary, shall include a description of its use.

'True nature' means a clear and accurate description of the characteristics of the food but it does not require a detailed description including all the main ingredients. The name should be easy to understand and should not be confused by superfluous adjectives.

As indicated above, a trade mark, brand name or fancy name may not be used as a substitute for a name satisfying the requirements of this regulation. However, it is acceptable to print a trade mark, brand name or fancy name in large type followed by a name which meets these requirements, provided always that the provisions as to intelligibility (see **6.4.23**) are met.

The Food Advisory Committee has recommended that the requirements as to food names should be clarified through non-statutory guidelines. No such guidelines have yet been considered but their introduction would not be subject to approval by the EC Commission. The Commission itself has moved in this direction by the publication of a Commission interpretive communication[6] on the names under which foodstuffs may be sold following the judgments of the European Court of Justice in the *Smanor*[7] and *Déserbais*[8] cases. Those cases concerned the names used by exporting member states which were not acceptable and may be confusing to consumers in the importing member state. The Commission offered opinions, not binding in law, as to the use of the names 'vinegar', 'yogurt' and 'caviar' and indicated its intention to issue further opinions in due course. This initiative by the Commission is controversial – the more so because one of the opinions is clearly wrong – on the grounds that only the courts can offer an authoritative interpretation of the law.

1 Food Labelling Regulations 1984, reg 7.
2 Food Labelling Regulations 1984, Sch 1.
3 Food Labelling Regulations 1984, reg 8.
4 Food Labelling Regulations 1984, reg 11.
5 Food Labelling Regulations 1984, reg 9.

6 Commission Communication (91/C 270/02) – OJ No C 270 of 15 October 1991.
7 Judgment of 14 July 1988 in Case 298/87 [1988] ECR 4489, ECJ.
8 Judgment of 22 September 1988 in Case 286/86 [1988] ECR 4907, ECJ.

Physical Condition or Treatment[1]

6.4.6 Where a food is powdered or is in any other physical condition or if it has been dried, freeze dried, frozen, concentrated, smoked or subjected to any other treatment, and the omission of an indication of such condition or treatment would be misleading to a purchaser, then the name of the food must be accompanied by such an indication. The following particular requirements apply:[2]

(1) If the food has been frozen with dichlorodifluoromethane the name must be accompanied by the words 'contact frozen with dichlorodifluoromethane';
(2) If meat, poultry meat or offal has been frozen and thawed the name must include or be accompanied by the words 'previously frozen – do not refreeze';
(3) Where meat has been treated with proteolytic enzymes or is derived from an animal which has been so treated it must be accompanied by the word 'tenderised';
(4) The name used for dried peas which have been soaked and then canned or frozen must be accompanied by the word 'dried', 'processed' or 'soaked'.

Apart from the four special cases mentioned above, whether or not a condition or treatment must be disclosed is a matter of common sense. It must not be assumed that a purchaser has special culinary skills which would render the statement unnecessary. The test is whether the ordinary consumer would be misled by its omission. In difficult matters, such as the correct manner to describe meat products which have been reformed or shaped and the declaration of mechanically recovered meat, the matter is usually resolved by agreement between trade bodies, the local authorities coordinating body on trading standards (LACOTS) and the Ministry of Agriculture, Fisheries and Food.

For the application of this requirement to irradiated ingredients see **6.4.14** below.

1 Food Labelling Regulations 1984, reg 12.
2 Food Labelling Regulations 1984, Sch 2.

List of Ingredients[1]

6.4.7 Where a food consists of two or more ingredients, a list headed by the word 'ingredients' must appear on the label listing all the ingredients in descending order by weight as determined at the time of use in the preparation of the food.

The order of ingredients in the list is determined at the 'mixing bowl' stage of production and not as they may be when a sample is taken of the finished product. It is sometimes the case that the order of ingredients determined by analysis of the finished product is different from that of the recipe because of chemical changes and the leaching of ingredients. Providing the ingredients are introduced to the product in accordance with the recipe and evidence can be produced to prove it, subsequent changes in the order may be disregarded. It should be borne in mind that enforcement officers have the power to examine recipes to verify the correctness of lists of ingredients.

If an ingredient is used in the making of the food in a concentrated or dehydrated form and is reconstituted during the preparation of the food, its position in the list of ingredients may be that based on the weight before concentration or dehydration.

Where the food itself is in concentrated or dehydrated form and is intended to be reconstituted before use by the addition of water the ingredients may be listed according to their weight when reconstituted provided the list of ingredients is accompanied by the words 'ingredients of the reconstituted product' or 'ingredients of the ready to use product' or by similar words.

Where a product contains mixed fruit, nuts, vegetables, spices or herbs and no particular fruit, nut, vegetable etc predominates they may be listed other than in descending order by weight as follows:

(a) if the food consists entirely of such a mixture the list of ingredients should be accompanied by the words 'in variable proportion' or similar words;
(b) if the food consists of such a mixture with other ingredients, that part of the list of ingredients which refers to the mixture should bear words similar to those in (a) above.

Added water must appear in the list of ingredients in its correct position unless it has been used in the preparation of the food solely for the reconstitution or partial reconstitution of an ingredient used in concentrated or dehydrated form, or if it is used as a medium not normally consumed, or if it does not exceed 5% of the finished product.

1 Food Labelling Regulations 1984, regs 13 and 14.

The Names of Ingredients[1]

6.4.8 The name used for an ingredient in the list of ingredients must be a name which could be lawfully used if the ingredient was being sold as a food in its own right. In most cases, therefore, the name used for an ingredient must be a specific name satisfying the requirements discussed in **6.4.5** above. However, certain generic names may be used as prescribed in the regulations[2] subject to certain conditions. They are:

Generic Name	*Ingredients*	*Conditions of use of Generic Name*
Caseinates	Any type of caseinate	
Cheese	Any type of cheese	The labelling of the food of which the cheese is an ingredient must not refer to a specific type of cheese
Citrus peel	Peel of any type of citrus fruit	
Cocoa butter	Press, expeller or refined cocoa butter	
Cream	Clotted cream, whipping cream, whipped cream, sterilised cream or single cream, sterilised half cream, and half cream	

Generic Name	*Ingredients*	*Conditions of use of Generic Name*
Half cream	Sterilised half cream and half cream	
Crumbs or rusks	Any type of crumbed, rusks baked cereal product	
Dextrose	Anhydrous dextrose or dextrose monohydrate	
Fat	Any refined fat or mixture or refined fats	Generic name must be accompanied by either: (a) the description 'animal' or 'vegetable' or both as appropriate; or (b) an indication of the specific animal origin or vegetable origin of the fat or both
Fish	Any species of fish or shell fish	The labelling of the food must not refer to a specific species of fish
Other fish	Any species of fish or shell fish other than one referred to in the labelling of the food	
Flour; followed by a list of the cereals from which the flour is derived in descending order by weight	Any mixture of flour derived from two or more types of cereal	
Crystallised fruit	Any crystallised fruit	Proportion of crystallised fruit in the food max 10%
Gum base	Any type of gum preparation used in the preparation of chewing gum	
Herbs, or mixed herbs	Any combination of two or more herbs or parts of herbs	Proportion of herbs in the food max 2%
Honey	Any type of honey*	
Meat	Any type of meat**	The labelling of the food must not refer to a specific type of meat
Other meat	Any type of meat product** other than a type referred to in the labelling of the food	
Offal	Any specified part of the carcase**	Generic name may only be used for a cooked food

Generic Name	*Ingredients*	*Conditions of use of Generic Name*
Poultry meat	Any type of poultry meat	The labelling of the food must not refer to a specific type of poultry meat
Nuts	Any nuts	Generic name may only be used if either: (a) the proportion of nuts in the food does not exceed 1%, or (b) the nuts are ingredients of a prepacked mixture of nuts with muscatels, raisins, sultanas, and/or currants and the net weight of the mixture is not more than 50g
Oil	Any refined oil, other than olive oil, or any mixture of refined oils not including olive oil	Generic name must be accompanied by either: (a) the description 'animal' or 'vegetable', or both as is appropriate or (b) an indication of the specific animal or vegetable origin of the oil, or both, as appropriate. In the case of hydrogenated oil the generic name must, unless it is accompanied only by the description 'animal' also bear the description 'hydrogenated'
Spices or mixed spices	Any combination of two or more spices and/or spice extracts	The proportion of spices and/or spice extracts in the food must not exceed 2%
Starch	Any starch other than one which has been modified chemically	
Modified starch	Any starch which has been modified chemically	
Sugar	Any type of sucrose	
Vine fruits	Any muscatels, raisins, sultanas or currants	

* Any type of honey as defined in the Honey Regulations 1976.

** Any type of meat as defined in the Meat Products and Spreadable Fish Products Regulations 1984 and, in the case of offal, any part of the carcase as specified in Part I or Part II of Sch 2 thereto.

The use of the above generic names for certain ingredients has been criticised with some justification by the Food Advisory Committee and the EC Commission is to review the matter soon.

1 Food Labelling Regulations 1984, reg 15 and Sch 3.
2 Food Labelling Regulations 1984, Sch 3.

Compound Ingredients[1]

6.4.9 Where an ingredient itself consists of two or more other ingredients it is known as a compound ingredient. Such compound ingredients may be named in the correct position in the list of ingredients as a single ingredient, but the constituents which make up the compound ingredient must themselves appear in a list either instead of or in addition to the compound ingredient. If the name of the compound ingredient is given, the names of its ingredients must follow that name immediately or appear in close proximity to it in such a way as to make it clear that they are ingredients of the compound ingredient.

The names of the ingredients of a compound ingredient need not be given in a case where –

(a) the compound ingredient would not be required to be marked or labelled with a list of ingredients if it were itself being sold as a food;
(b) the compound ingredient is identified in the list of ingredients by a generic name (see **6.4.8**);
(c) the compound ingredient constitutes less than 25% of the finished product.

The Food Advisory Committee has recommended that the exemption from ingredient listing for a compound ingredient constituting less than 25% of the finished product should be abolished. It is believed that this view is shared by the EC Commission.

1 Food Labelling Regulations 1984, reg 16.

Ingredients which need not be named[1]

6.4.10 Ingredients need not be named in the list of ingredients in the following circumstances:

(a) where the constituents of an ingredient have become separated during a manufacturing process and are later re-introduced in their original proportions;
(b) where an additive is used in an ingredient but serves no significant technological function in the finished product;
(c) where an additive is used solely as a processing aid;
(d) any substance other than water which is used as a solvent or carrier for an additive and is used in an amount no greater than that which is strictly necessary for the purpose.

1 Food Labelling Regulations 1984, reg 18.

Foods which need not bear a List of Ingredients[1]

6.4.11 The following foods are not required to bear a list of ingredients:

(1) fresh fruit and vegetables, including potatoes which have not been peeled or cut into pieces;
(2) carbonated water, to which no ingredient other than carbon dioxide has been added and whose name indicates that it has been carbonated;
(3) vinegar which is derived by fermentation exclusively from a single basic product and to which no other ingredient has been added;*
(4) cheese, butter, fermented milk and fermented cream to which no ingredient has been added other than lactic products, enzymes and micro-organism cultures essential to manufacture or, in the case of cheese other than fresh curd cheese and processed cheese, such amount of salt as is needed for its manufacture;*
(5) flavourings;
(6) food consisting of a single ingredient, including flour, to which no substances have been added other than those which are required to be added;
(7) any drink with an alcoholic strength by volume of more than 1.2%.

* In these cases only other added ingredients need be listed but the list must be headed by the words 'added ingredients'.

If a list of ingredients is given in respect of any exempted food it must be a complete list in accordance with the Labelling Regulations.

The Food Advisory Committee has recommended that all exemptions from ingredient listing should be reviewed – especially alcoholic drinks.

1 Food Labelling Regulations 1984, reg 19.

Special Emphasis of Ingredients[1]

6.4.12 Where a food is characterised by the presence of a particular ingredient or by the low content of a particular ingredient, the labelling of the food must not place special emphasis on the presence or low content of that ingredient, unless it includes a declaration of the minimum or maximum percentage, as appropriate, of that ingredient in the food. The declaration must appear either next to the name of the food or in the list of ingredients in close proximity to the name of the ingredient in question.

A reference in the name of a food to a particular ingredient, or to an ingredient which is used in a small quantity and only as a flavouring, shall not of itself constitute the placing of special emphasis on the presence or low content of that ingredient.

What constitutes characterisation is not defined and nor is the meaning of 'special emphasis'. It is clear that a food may be characterised by an ingredient giving a distinctive taste or flavour but other attributes of an ingredient may also qualify. Special emphasis may be given by the reference to an ingredient in the product name in larger print than the rest of the name or in a distinctive colour or by pictorial means.

It should be noted that some regulations dealing with particular food types require a declaration of the minimum content of the major ingredient (see **6.4.21**). Compliance with those regulations may also provide an opportunity to give special emphasis to that ingredient.

1 Food Labelling Regulations 1984, reg 20.

Declaration of Additives[1]

6.4.13 Additives must be declared in the list of ingredients in the same way as any other ingredient, but any additive serving the technological function of any of the following categories must be identified by the category name followed by its specific name and/or serial number. An additive which does not serve a technological function of one of the following categories must be declared in the list of ingredients by its specific name:

Acids*
Acidity regulators
Anti-caking agents
Anti-foaming agents
Antioxidants
Artificial sweeteners
Colours
Emulsifiers
Emulsifying salts***
Flavour enhancers
Flavourings**
Flavour improvers
Gelling agents
Glazing agents
Preservatives
Raising agents
Stabilisers
Thickeners

* In the case of an additive which is added to or used in food to serve the function of an acid and whose specific name includes the word 'acid', it is not necessary to use the category name.
** For the time being it is sufficient to use the category name 'flavourings' alone.
*** This category name may be used only for an emulsifying salt, as defined in the Cheese Regulations 1970, which is added to or used in processed cheese (including cheese spread).

For a discussion of the control system for additives see Chapter 9. For approved lists of additives see Appendix C.

1 Food Labelling Regulations 1984, reg 15 (4), (5), (6) and Sch 4.

Indication of Irradiation[1]

6.4.14 Food which has been irradiated or which contains irradiated ingredients must, subject to certain exemptions, be marked 'irradiated' or 'treated with ionising irradiation'. Not pre-packed food, food prepacked for direct sale, certain flour confectionery and fancy confectionery products need not be so marked if they are not exposed for sale.

1 Food Labelling Regulations 1984, regs 12, 15 and 26A.

'Use by' Dates[1]

6.4.15 Foods which, from the microbiological point of view, are highly perishable and in consequence likely after a short period to constitute an immediate danger to human health, must be marked with a 'use by' date. The 'use by' date must be indicated by the words 'use by . . .' followed by the date up to and including which the food, if properly stored, is recommended for use, and any storage conditions which need to be observed. The actual date may appear separately from the words 'use by' provided that those words are followed by a reference to the place where the date appears, eg 'see lid'.

The decision as to whether a particular food requires a 'use by' date is one for the manufacturer or packer who originally marks it. However, the Ministry of Agriculture, Fisheries and Food has issued *Guidelines on Use By Dates* in which it is suggested that the foods which should be so marked are those which should be maintained at low temperatures of up to 8 degrees C for their safety, eg prepared desserts, prepared salads, ready-made meals, cooked meats, pate and soft cheese, together with those in which spoilage bacteria or pathogens might multiply to high numbers, or are likely to cross-contaminate other foods, eg fresh meat, fish and poultry. Bread and cakes which deteriorate quickly do not pose a danger to health and need not be marked with a 'use by' date. Useful additional guidance as to the foods which ought to be marked with a 'use by' date is contained in the Food Hygiene (Amendment) Regulations 1990 and guidelines thereon issued by the Department of Health – see Chapter 13. For exempted foods see **6.4.16** below.

It has become the practice in certain large retailing establishments to mark food which does not fall within the criteria discussed above with a use by date. For operational reasons it is convenient to have all foods in chilled displays. This practice is not unlawful but it is contrary to the spirit of the regulations and may be confusing for consumers.

It is an offence to sell any food after the date shown in a use by date and for any person other than the manufacturer, packer or seller originally responsible for marking the 'use by' date to alter that date.[2]

1 Food Labelling Regulations 1984, regs 21A and 22.
2 Food Labelling Regulations 1984, reg 40(f) and (g).

Minimum Durability Dates[1]

6.4.16 Subject to exemptions (see **6.4.17**) foods which do not bear a 'use by' date must be marked with a 'best before' date in the form 'best before' followed by the date up to and including which the food can reasonably be expected to retain its specific properties if properly stored and any storage conditions which need to be observed if the food is to retain its specific properties up to that date.

In the case of a food which can reasonably be expected to retain its specific properties for –

(1) three months or less the date may be expressed in terms of a day and month only eg 'Best before 15 October':
(2) more than three months but not more than 18 months the date may be expressed in terms of a month and year only eg 'Best before end October 1992':
(3) more than 18 months the date may be expressed either in terms of a month and year only or in terms of a year only eg 'Best before end December 1992' or 'Best before end 1992'.

The date may appear on the labelling separately from the words 'best before' or 'best before end' if it is followed by a reference to the place where the date appears eg 'best before end – see top of can'.

It is an offence for any person other than the manufacturer, packer or seller originally responsible for marking the food to alter a best before date. Unlike 'use by' dates it is not an offence of itself to sell food after the expiry of a best before date but if the food is not of the substance or quality demanded or if it fails the food safety requirement (see Chapter 10) the fact that the best before date had expired would be damaging to the defendant. It may also be significant in any civil proceedings which may be brought by an aggrieved purchaser – see Chapter 2.

The following foods need not be labelled or marked with an indication of minimum durability or a 'use by' date:

(a) fresh fruit and vegetables (including potatoes but not including sprouting seeds, legume sprouts and similar products) which have not been peeled or cut into pieces;
(b) wine, liqueur wine, sparkling wine, aromatised wine and any similar drink obtained from fruit other than grapes and any drink made from grapes or grape musts and coming within codes 2206 00 91, 2206 00 93 and 2206 00 99 of the Combined Nomenclature;[2]
(c) any drink with an alcoholic strength by volume of 10% or more;
(d) any soft drink, fruit juice or fruit nectar or alcoholic drink, sold in a container containing more than five litres and intended for supply to catering establishments;
(e) flour confectionery and bread which are normally consumed within 24 hours of their preparation;
(f) vinegar;
(g) cooking and table salt;
(h) solid sugar and products consisting almost solely of flavoured or coloured sugars;
(i) chewing gum and similar products;
(j) edible ices in individual portions;
(k) until 20 June 1992 –

(i) deep frozen food,
(ii) cheese which is intended to ripen completely or partially in its packaging,
(iii) any food which, if properly stored, can be expected to retain its specific properties for more than 18 months.

The practice of certain large retailers to mark exempted foods with indications such as 'display until' or 'do not sell after' dates may, on a strict interpretation of the regulations, be unlawful. However, it is an unobjectionable practice from the consumer's point of view.

The Food Advisory Committee has suggested that where strict storage temperatures are required to maintain food safety and quality the appropriate temperatures should be given on the label.

1 Food Labelling Regulations 1984, regs 21 and 22.
2 Council Regulation (EEC) No 2658/87 as amended (OJ No L 259; 16.09.91).

Lot or Batch Marking

6.4.17 Regulations are imminent to implement Council Directive 89/396/EEC on lot marking. The regulations will create a 'lot marking obligation' by which food, other than exempted food, will have to be marked with a lot indication if it comprises:

(i) a production sales unit, that is to say a sales unit of food produced as part of a batch of such sales units produced under similar conditions;
(ii) a preparation sales unit, that is to say a sales unit of food prepared as part of a batch of such sales units prepared under similar conditions; or
(iii) a packaging sales unit, that is to say a sales unit of food packaged as part of a batch of sales units packaged under similar conditions.

A 'sales unit' of food comprises food in the process of being, or which can be reasonably expected to be, sold as an identifiable unit.

Lot marking indications will have to appear on the prepackaging or on a label attached to the prepackaging on sales units or, if the sales unit is not prepacked, on a container in which the sales unit is contained or on a commercial document accompanying the sales unit. The lot marking indication must be easily visible, clearly legible and indelible.

The lot marking indication of a sales unit is an indication determined by the lot marking determiner (ie manufacturer, or importer) which is preceded by the letter 'L' except in cases where that indication is clearly distinguishable from other indications relating to the sales unit and identifies the appropriate batch within which the sales unit falls.

There will be exemptions for products marked with a 'use by' date or a minimum durability indication; for agricultural produce; for non-prepacked foods or foods packed by the retailer for direct sale; for certain small packs and for certain prepacked foods sold in catering establishments.

Note: The above summary has been prepared from consultative documents and should not, therefore, be relied upon.

Instructions for Use[1]

6.4.18 Where instructions for the use of foods are given they must be such as to enable appropriate use to be made of the food. Where a substance other than water must be added to a dry mix food the substance to be used must be specified in immediate proximity to the name of the food. In any advertisement, whether written or oral, for the sale of a concentrated, dry mix or similar food, any substance, other than water, which must be added must be indicated.

1 Food Labelling Regulations 1984, reg 23.

Name and Address[1]

6.4.19 The name and address (see **6.4.4** (e)) must be sufficient for a letter addressed to it to arrive.

1 Food Labelling Regulations 1984, reg 6.

Origin Marking[1]

6.4.20 The origin marking requirement (see **6.4.4** (f)) is required where any indication or pictorial representation might mislead a consumer as to the true place of origin of the food. Thus a product manufactured in London and named 'Ardennes Pate' would have to be marked 'made in London' or 'Ardennes Style Pate'.

There is no general requirement for origin marking and the former requirements of the Trade Descriptions Act 1972 have been repealed. However, the EC Commission has proposed a directive to require that products with a particular geographic provenance should be registered as such and the production of competing products not conforming to the specification in the register should be prohibited. It is a controversial proposal which may not come to pass.

It is an offence against the Trade Descriptions Act 1968[2] to give a false or misleading indication of origin. For the purposes of that Act the country of origin of a product is that country in which it last underwent a treatment or process resulting in a substantial change. There is no offence in the Food Safety Act 1990 to cover false or misleading indications of origin.

Where foods are imported in bulk and prepacked in the United Kingdom, or where a food is imported and then subject to further processing, it is sometimes difficult to decide on the true country of origin. Given current concern about origin marking it is wise to interpret the requirements of the Food Labelling Regulations on this point widely. If in doubt, an indication of the true place of origin should be given.

The Food Advisory Committee has recommended that there should be non-statutory guidelines on the origin marking of food.

No such guidelines have yet been produced.

1 Food Labelling Regulations 1984, reg 6.
2 Trade Descriptions Act 1968, ss 1 and 36.

Additional Labelling Requirements for Certain Categories of Food[1]

6.4.21 Certain additional labelling requirements are imposed for food sold from vending machines and certain alcoholic drinks. They are:

(a) Where food is sold from a vending machine the name of the food must appear on the front of the machine unless the name on the labelling of the food can be easily and clearly seen through the outside of the machine.

(b) Prepacked alcoholic drinks with an alcoholic strength by volume of more than 1.2%, other than Community-controlled wine, must be marked or labelled with an indication of the alcoholic strength by volume in the form of a figure to not more than one decimal place in one of the following forms –

'10.5% vol', or 'alcohol 10.5% vol' or 'alc 10.5% vol'.

Positive and negative tolerances for the indication are prescribed.[2] The alcoholic strength must be determined at 20°C.

Note: Consequent upon the Licensing (Low Alcohol Drinks) Act 1990, which requires a licence for the sale of drinks with an alcoholic strength above 0.5% vol, the above strength marking requirements will soon be amended to require declarations of alcoholic strength for all drinks above that figure.

(c) Provision is made for whisky with an alcoholic strength of less than 40% vol, brandy, gin, rum or vodka with an alcoholic strength of less than 37.2% vol (other than brandy whose alcoholic strength has fallen to less than 37.2% vol only through maturing in cask) to be marked or labelled with the words 'under strength'. However, by Council Regulation (EEC) No 1576/89,[3] the sale of whisky/whiskey below 40% vol, rum, gin and vodka below 37.5% vol and brandy below 36% vol became unlawful from 15 December 1989.

(d) For alcoholic drinks, including Community-controlled wine, with a strength of more than 1.2% which are sold otherwise than prepacked, the alcoholic strength of a representative sample must be displayed as indicated in (b) above. The representative sample need not exceed 30 save for Community-controlled wine where it need not exceed six. Positive and negative tolerances are applied.[2] This requirement does not apply to cocktails or other drinks made up of mixtures or more than one drink nor to drinks where it is customary to serve in such a way that information in respect of the alcoholic strength can readily be seen by the intending purchaser.

1 Food Labelling Regulations 1984, regs 30, 31 and 31A.
2 Food Labelling Regulations 1984, Sch 5A.
3 Council Regulation (EEC) No 1576/89 given effect by the Spirit Drinks Regulations 1990/1179.

Food which is not Prepacked etc[1]

6.4.22 Food which is not prepacked, which is intended for direct sale, certain flour confectionery, certain fancy confectionery products, and food sold at catering establishments is not required to be labelled or marked with any of the statutory information described in this chapter except the name of the food and, when appropriate, an indication that it has been irradiated. However, where any additives serving the function of an antioxidant, artificial sweetener, colour, flavour enhancer, flavouring or preservative has been used they must be marked or labelled with an indication of such categories of additives, eg 'Orange flavour bon bons – with colour, preservative and flavour'.

'Direct sale' means food prepacked by a retailer for sale by him on the premises where the food is packed or from a vehicle or stall used by him, and in relation to edible ices, flour confectionery or bread, prepacked by a retailer for sale as above or prepacked for sale on premises where business is conducted by the producer of the food under the same name as the business conducted on the premises where the food is produced.

1 Food Labelling Regulations 1984, regs 24, 25, 26 and 28.

Labelling and Marking[1]

6.4.23 The regulations provide that all the information required to be marked or labelled on food etc shall appear on the packaging; on a label attached to the packaging; on a label that is clearly visible through the packaging; in relevant trade documents; on lists or menus, as appropriate. All such information must be easy to understand, clearly legible and indelible. The name of the food, the indication of minimum durability or a 'use by' date, alcoholic strength and indication of quantity, as required, must appear in the same field of vision. The term 'same field of vision' means simultaneously visible under normal conditions of purchase.

For nutrition labelling, claims, misleading descriptions and a summary of the labelling requirements for special category foods see Chapter 7.

1 Food Labelling Regulations 1984, regs 32, 33, 34 and 35.

Chapter 7

Misleading Descriptions and Special Category Labelling

7.1 Misleading Descriptions

General Provisions

7.1.1 Any description of a food product which is misleading may be an offence against the Food Safety Act 1990.[1] Whether or not a description is misleading is a question of fact to be determined in each case. In addition to the general offences in the Food Safety Act 1990, the Food Labelling Regulations 1984 prohibit the use of certain words and descriptions[2] in the labelling or advertising of a food unless certain conditions are satisfied.

1 Food Safety Act 1990, ss 14 and 15.
2 Food Labelling Regulations 1984/1305, reg 38 and Sch 7.

Controlled Descriptions

7.1.2 The following are the principal descriptions which are subject to conditions imposed by the Food Labelling Regulations 1984 together, where appropriate, with notes as to other legislation affecting the use of such descriptions.

7.1.3 *Butter*. The word 'butter' or any other word or description which implies that the food being described contains butter may not be applied to any chocolate confectionery or sugar confectionery or to any part of any chocolate confectionery or sugar confectionery, unless at least 4% of the confectionery or part to which the word or description is applied consists of milk fat. The word may not be used in the name of any biscuit unless at least half the fat used in the preparation of the biscuit is milk fat, and at least 7% of the biscuit consists of milk fat.

The word 'butter' is a reserved description which may not be used in the labelling of a food unless the product conforms to prescribed compositional standards (**5.3.2**). Further, claims that margarine contains butter or resembles butter are controlled[1] and the use of the word 'butter' for non-dairy based products is prohibited by EC Regulations.[2]

1 Margarine Regulations 1967/1867, reg 8.
2 Council Regulation (EEC) No 1898/87; Milk and Milk Products (Protection of Designations) Regulations 1990/607.

7.1.4 *Cream.* The word 'cream' or any other word or description which implies that the food being described contains cream may not be applied to any chocolate confectionery or sugar confectionery or to a part of any chocolate confectionery or sugar confectionery, unless at least 4% of the confectionery or part to which the word or description is applied consists of milk fat.

The word 'cream', togther with derivatives and variants, is a name prescribed by law and may not be used for any product which does not satisfy prescribed compositional standards.[1] It is also subject to the EC Regulation cited at **7.1.3** above.

1 The Cream Regulations 1970/752, regs 4 and 7.

7.1.5 *Dietary or dietetic.* The descriptions 'dietary' or 'dietetic'may not be applied to any food unless the food has been specially made for a class of persons whose digestive process or metabolism is disturbed or who, by reason of their special physiological condition, obtain special benefit from a controlled consumption of certain substances and the food is suitable for fulfilling the particular nutritional requirements of that class of persons.

7.1.6 *Flavours.* Any description incorporating the name of a food in such a way as to imply that the food, or the part of the food, being described has the flavour of the food named in the description may not be applied to any food unless the flavour of the food is derived wholly or mainly from the food named in the description. However, any description incorporating the word 'chocolate' which is such as to imply that the food has a chocolate flavour may be applied to a food which has a chocolate flavour derived wholly or mainly from non-fat cocoa solids.

The foregoing does not prevent the use of the word 'flavour' preceded by the name of a food when the flavour of the food is not derived wholly or mainly from the food named in the description. This is to allow the use of artificial flavours and to distinguish them from flavours derived from natural foods.

A pictorial representation of a food which is such as to imply that the food to which the representation is applied has the flavour of the food depicted in the representation may not be applied to any food unless the flavour of the food to which the representation is applied is derived wholly or mainly from the depicted food.

The above provisions as to flavour do not prohibit the use of any word or pictorial device suggestive of fruit or fruits for a soft drink, as defined in the Soft Drinks Regulations 1964[1] if such use is part of a customary name.[2]

New EC provisions made under the framework directive on flavourings[3] expected shortly may change the above requirements.

1 The Soft Drinks Regulations 1964/760, reg 2.
2 See **6.4.5** (ii).
3 Council Directive 88/388/EEC (OJ No L 184; 15.07.88).

7.1.7 *Fresh, Garden or Green.* These descriptions may not be applied to any canned or frozen peas which have been dried and soaked before being canned or frozen.

7.1.8 *Milk.* The word 'milk' or any other word or description which implies that the food being described contains milk may not be used as part of the name of a food unless –

(a) the food has as an ingredient cow's milk with all its normal constituents in their natural proportions; or

(b) (i) the food has as an ingredient cow's milk which has been subjected to a process of treatment, and
 (ii) the word or description is accompanied by an indication of that process or treatment; or

(c) (i) the food has as an ingredient the milk of an animal other than the cow with all its normal constituents in their natural proportions, and
 (ii) the word or description is accompanied by the name of the animal; or

(d) (i) the food has as an ingredient the milk of an animal other than the cow which has been subjected to a process or treatment, and
 (ii) the word or description is accompanied by the name of that animal and an indication of that process or treatment; or

(e) the word or description is used in such a way as to indicate clearly that it does not refer to the presence of milk or any of its constituents in the food, or

(f) the word or description is used in accordance with any regulations or Order made or continued in force under the Food Safety Act 1990.

The word 'milk' may not be used as the name of an ingredient in food unless it is cow's milk or the word is accompanied by the name of the animal from which the milk has been obtained and its use complies in all other respects with the Food Labelling Regulations 1984.

Quality standards are laid down for milk[1] and if the milk being sold is not milk of the bovine species its origin must be stated.[2] The word 'milk' and certain derivatives and variants are names required by law[3] and compositional standards must be satisfied.

1 Council Regulation (EEC) No 1411/71 (OJ No L148; 03.07.71), art 3 (1).
2 Council Regulation (EEC) No 1898/87 (OJ No L 182; OJ No L 03.07.87).
3 Drinking Milk Regulations 1976/1883.

7.1.9 *Starch reduced.* This description may not be applied to any food unless less than 50% of the food consists of anhydrous carbohydrate calculated by weight on the dry matter of the food, and the starch content of a given quantity of the food is substantially less than that of the same quantity of similar foods to which the description is not applied.

7.1.10 *Alcoholic drinks.* A number of descriptions used in relation to alcoholic drinks or claims as to the amount of alcohol are controlled. They should be considered together with the requirements for the declaration of alcoholic strength (**6.4.21**) and EC marketing standards for spirit drinks.[1] The requirements are:

(a) *Alcohol free* may not be applied to any alcoholic drinks from which the alcohol has been extracted unless the drink has a strength of not more than 0.05% vol and it is marked or labelled with an indication of its maximum strength immediately preceded by the words 'not more than' or, in appropriate case, with an indication that it contains no alcohol.

(b) *Dealcoholised* may not be applied to any drink unless the drink, being an alcoholic drink from which the alcohol has been extracted, has an alcoholic strength of not more than 0.5% vol, and is marked or labelled with an indication of its maximum alcoholic strength immediately preceded by the words 'not more than' or, in an appropriate case, with an indication that it contains no alcohol.

(c) *Low Alcohol* or any other word or description which implies that the drink being described is low in alcohol may not be applied to an alcoholic drink unless the drink has an alcoholic strength of not more than 1.2% and the drink is marked or labelled with an indication of its maximum alcoholic strength immediately preceded by the words 'not more than'.

(d) *Non-alcoholic* may not be used in conjuction with a name commonly associated with an alcoholic drink, except in the composite name 'non-alcoholic wine' when that composite name is used as discussed in **7.2** below.

(e) *Shandy type products* Restriction on the use of such terms as 'shandy', 'shandygaff', 'cider shandy' and 'cider shandygaff' linked to a minimum alcoholic strength of 0.9% vol will shortly be revoked. The Licensing (Low Alcohol Drinks) Act 1990 requires that drinks containing above 0.5% vol of alcohol should be licensed. The new provisions will therefore require that all drinks with an alcoholic strength of more than 0.5% vol will have to bear a declaration of alcoholic strength.

(f) *The words 'sweetened liqueur'* may not be applied to any drink except a suitably flavoured compound spirit which has been rendered sweet and viscous only by the addition of carbohydrate sweetening matter.

(g) *The name 'tonic wine'* may not be applied to any drink unless there appears in immediate proximity to the words 'tonic wine' the clear statement 'the name 'tonic wine' does not imply health giving or medicinal properties'. No recommendation as to consumption or dosage may appear in the labelling or advertising of the drink.

(h) *The description 'vintage'* may not be applied to any drink except brandy; or cider or perry which is the product of the fresh juice of apples or pears, as the case may be, harvested in the designated vintage year and which is marked or labelled with an indication of that year; or liqueur wine or wine which, in either case, is obtained by the fermentation in its district of origin of the juice of fresh grapes.

(i) *The words 'Scotch whisky', 'Irish whiskey', 'blended Scotch whisky' or 'blended Irish whiskey'* may not be used as the name of the spirit unless the spirit is as named and defined.[2] Note, however, the complementary provisions of the Scotch Whisky Act 1988 and the Scotch Whisky Order 1990, see **5.3.2**.

1 Council Regulation (EEC) No 1576/89 (OJ No L 160; 12.06.89); the Spirit Drinks Regulations 1990/1179.
2 The definition is in para 1 of Sch 7 to the Finance Act 1969.

7.2 Wine

7.2.1 The marketing of wine is controlled by a series of ever-changing and very complex EC Regulations – see Appendix 2. These regulations are given the force of law in the United Kingdom by the Common Agricultural Policy (Wine) Regulations 1990[1] and are enforced at the importation stage by Customs and Excise, at all subsequent stages other than retail by the Wine Standards Board[2] and at retail level by local food authorities. The production, bottling, distribution, labelling and advertising of wine is very closely controlled and is seldom a cause for concern. Accordingly only a brief summary of the labelling requirements[3] is appropriate in this book.

1 The Common Agricultural Policy (Wine) Regulations 1990/1363.
2 The Wine Standards Board, Five Kings House, 1 Queen Street Place, London EC4R 1QS; tel: 071–236–9512.
3 Council Regulation (EEC) No 2392/89 (OJ No L 232; 09.08.89).

7.2.2 *Table Wines Originating within the EC*

Mandatory Information

The words 'Table Wine'.

The nominal volume.

For containers of not more than 60 litres – the name or business name of the bottler, the local administrative area and the member state.

For other containers – the name or business name of the consignor, the local administrative area and the member state.

In the case of a consignment to another member state or export: the name of the member state in whose territory the grapes were harvested and turned into wine provided both operations took place in the same member state.

In the case of a table wine made in a member state other than that in which the grapes were harvested – the words 'wine made in . . . from grapes harvested in . . .'.

Table wine resulting from a mixture of grapes or from coupage or products originating in more than one member state the words 'blend of wines from different countries of the EC'.

For certain designated wines the word 'Retsina'.

In the case of wines obtained in Spain by mixing red wines with white wines: the words 'vino tinto mezela' in that Spanish territory.

The actual alcoholic strength, the natural or technical conditions, and the ageing of the wine.

Permitted Information

A statement of whether the wine is red, rosé or white or, in the case of Spain, a mixture thereof.

A brand name.

A distributor's name and address and their local administrative area.

Any distinction awarded.

Where the wine is not exported or consigned to another member state or exported – the name of the member state where the grapes were harvested and turned into wine.

Certain analytical data.

Recommendations to the consumer as to use.

The 'e' mark.[1]

A geographical unit which is smaller than a member state.

The name of one or two vine varieties.

A vintage year.

An award granted by an the official body.

Where the wine was bottled at a vineyard a statement of that vineyard.

Information as to the history of the wine.

7.2.3 *Quality wines produced in specified regions*

Mandatory Information

Name of the specified region of origin.

Quality wine PSR of equivalent expression (appellation controlée) no abbreviations.

Nominal volume.

For containers of not more than 60 litres: the name or business name of the bottler and the local administrative area and member state.

For other containers: the name and business name of the consignor; the local administrative area and the member state.

In the case of a consignment to another member state or export: the name of the member state in which the specified region is situated.

The actual alcoholic strength.

Permitted Information

Whether the wine is red, white or rosé.

The vintage year.

A brand name.

A distributor's name, and local administrative area.

Any distinction awarded.

The member state of origin.

Certain analytical data.

A recommendation as to use.

Additional details of a traditional kind.

Details as to the method of production, type of product and the colour.

Name of a geographical unit smaller than a member name of a vineyard or group of vineyards.

The names of one or two vine varieties.

A quality control number.

Where bottled.

Container or lot number.

History of the wine, natural or technical conditions of production and ageing.

The 'e' mark.[1]

1 Mark for packages satisfying Council Directive 75/106/EEC on the average quantity system.

7.2.4 *Products other than Table or Quality Wines PSR*

Mandatory Information

Type of product using prescribed definitions or traditional or customary names for products sold only in the originating member state.

For grape must or concentrated grape must – the density.

For other wine – the actual and total alcoholic strength.

Nominal volume.

For containers of more than 60 litres: the

Permitted Information

The vintage year.

Names or business names of distributors and local administrative area.

Where the product is not consigned to another member state or exported: the name of the member state where the grapes were harvested and production took place.

The 'e' mark.[1]

Analytical and other data.

Mandatory Information

name or business name of the bottler, the local adminstrative area and member state.

For other containers: the name or business name of the consignor, the local administrative area and member state.

For consignment to another member state or export: for wines – the member state where the grapes were harvested and made into wine; for grape musts – the member state where the grapes were harvested and production was carried out.

For wines and grape musts from a coupage of products from more than one member state: the words 'made from products of different countries of the EC'.

For grape musts not produced and wine not made in the member state where the grapes were harvested: the initials 'EEC'.

Any restriction on use imposed by EC legislation.

Permitted Information

For fermented grape must for direct human consumption: name of geographical unit, name of a vine variety and whether the product is red, white or rosé.

1 Mark for packages satisfying Council Directive 75/106/EEC on the average quantity system.

7.2.5 Products Originating in Third Countries

Mandatory Information

The word 'wine'.

The nominal volume.

Wine in containers of not more than 60 litres: name or business name of bottler or importer, administrative area and if bottled in a member state, the name of that member state.

In other containers: the name or business name of the importer, administrative area and member state.

The third country of origin.

Actual alcoholic strength.

Permitted Information

Whether the wine is red, white or rosé.

A brand name.

Name or business names of importers and local administrative area.

Analytical data.

Recommendations as to use.

Information concerning the history, natural conditions and ageing of the wine.

Any distinction awarded.

The 'e' mark.[1]

1 Mark for packages satisfying Council Directive 75/106/EEC on the average quantity system.

7.2.6 *Sparkling Wines and Aerated Sparkling Wines*[1]

Mandatory Information	*Permitted Information*
Name of product to include specified descriptions and the words 'sparkling wine'.	Other particulars may be given provided they are not liable to mislead.
Nominal volume.	
Type of product.	
Actual alcoholic strength.	
Name or business name of the producer or vendor in the EC.	
Administrative area and member state.	
Name or business name of the importer, local administrative area and member state.	
Quality wine PSR or equivalent expression.	
Name of region in which grapes were harvested.	
If produced from imported wines, a statement to that effect and the name of the third country.	
Name of vine variety for quality wines or the words 'produced from aromatic varieties of grape'.	

1 Council Regulation (EEC) No 3309/85 (OJ No L 320; 29.11.85).

Aromatised Wines, Aromatised Wine-based Drinks and Aromatised Wine Product Cocktails[1]

7.2.7 Definitions and sales descriptions for these products are laid down. Detailed labelling rules are pending.

1 Council Regulation (EEC) No 1601/91 (OJ No L 149; 14.06.91).

Use of Word 'Wine' in Composite Names for Foods or Drinks which are not Wine[1]

7.2.8 The appellation 'wine' is restricted by EC regulations to products which satisfy the definitions of wine contained therein. However, the word 'wine' may be used in a composite name in the labelling or advertising of food for a drink which is derived wholly or partly from fruit other than grapes or from other vegetable, plant or carbohydrate material, providing it is immediately preceded in the composite name by an indication of that fruit, vegetable, plant or carbohydrate material.

When the word 'wine' is used in a composite name for a drink which is derived wholly or partly from a mixture of such ingredients as are mentioned above and which is characterised by one or more of those ingredients, it is sufficient to specify such of those ingredients as characterise the drink.

The composite name 'non-alcoholic wine' may not be used except for a drink derived from unfermented grape juice which is intended exclusively for sacramental use and is labelled as such.

Each word in a composite name complying with the above requirements must appear in lettering of the same type and colour and of such a height that the composite name is clearly distinguishable from other products.

1 Food Labelling Regulations 1984, reg 39.

7.3 Eggs

7.3.1 *The marketing of eggs* is controlled by EC regulations[1] given the force of law in the United Kingdom by regulations continued in force under the Food Safety Act 1990.[2] The regulations lay down quality standards for eggs, grading standards by weight and labelling requirements. The regulations are enforced by trading standards officers of local food authorities.

1 Council Regulation (EEC) No 1907/90 (OJ No L 173; 06.07.90).
2 The Eggs (Marketing Standards) Regulations 1985/1271.

7.3.2 *Quality Standards* for eggs are –

- Class A or 'fresh eggs',
- Class B or 'second quality or preserved eggs',
- Class C or 'down graded eggs intended for the food industry'.

The quality standards for Grade A and B eggs are based on the shape and condition of the shell and cuticle; the air space; the clarity and consistency of the white; the visibility on candling of the yolk and its freedom from extraneous matter; the absence of germ cell; and freedom from odour. Grade C eggs are eggs other than Grade A or B eggs.

7.3.3 *Weight Grades* for Class A eggs are laid down. They are –

Grade 0: 75 g and over,
Grade 1: from 70 g up to 75 g,
Grade 2: from 65 g up to 70 g,
Grade 3: from 60 g up to 65 g;
Grade 4: from 55 g up to 60 g;
Grade 5: from 50 g up to 55 g;
Grade 6: from 45 g up to 50 g;
Grade 7: under 45 g.

7.3.4 *Labelling* of small packs of eggs for retail sale includes the following mandatory and optional information:

Mandatory Information	*Optional Information*
Name or business name and address of the undertaking who packed the eggs or	A selling price.

Mandatory Information	*Optional Information*
had them packed; the name business name or trade mark used by that undertaking, which may be a trade mark used collectively by a number of undertakings, may be shown if it contains no wording incompatible with the regulation as to the quality or freshness of the eggs, the type of farming used, or the origin of the eggs.	Retail management stock control codes.
Distinguishing number of the packing centre in its prescribed form.	A sell by or best before date.
Quality and weight gradings. Class Grade A eggs may be identified either by the words 'Class A' or the the letter 'A' whether alone or in combination with the word 'fresh'.	Particulars as to special storage conditions.
The number of eggs packed.	Statements designed to increase sales providing they are not misleading.
The packing date (in prescribed form).	Grade A eggs may be marked with one of the following: 'Free range eggs' 'Semi-intensive eggs' 'Deep litter eggs' 'Perchery (barn) eggs'.
Particulars as to refrigeration or to the method of preservation, in uncoded form, in respect of refrigerated or preserved eggs.	
Eggs from third countries: In addition to the above:	
The country of origin.	
Name of the packer in the third country.	

7.4 Fish

7.4.1 *Marketing Standards* for fresh and chilled fish, shrimps, edible crabs and lobsters, certain fishery products and preserved sardines are prescribed by EC regulations.[1] With the exception of preserved sardines[2] these regulations have not been given the force of law in the United Kingdom because they concern the quality of fishery products only and do not require additional labelling.

1 Council Regulations (EEC) Nos 3796/81 (OJ No L 379; 31.12.81); 103/76 (OJ No L 020; 28.01.76); 104/76 (OJ No L 020; 28.01.76); 2136/89 (OJ No L 212; 22.07.89).
2 Preserved Sardines (Marketing Standards) Regulations 1990/1084.

7.4.2 *Preserved Sardines.* The marketing standards lay down definitions for preserved sardines; the quality of such fish; the covering medium which may be used and appropriate sterilisation.

7.5 Fruit and Vegetables

7.5.1 *Marketing Standards* for fresh produce are prescribed by EC regulations[1] for apples and pears, apricots, artichokes, asparagus, beans, brussels sprouts, cabbages, carrots, cauliflowers, celery, cherries, chicory, cucumbers, garlic, grapes, kiwifruit, leeks, lettuce and endives, nectarines, onions, oranges and other citrus fruit, peaches, peas, plums, spinach, strawberries, sweet peppers, and tomatoes. Quality standards for each concerning their colour, cleanliness, shape and size are laid down together with the markings which must appear on wholesale packaging. The regulations are enforced by HM Horticultural Produce Marketing Inspectorate.

Quality standards apply at all stages of distribution including imports and exports. The following transactions are, however, exempt:

(a) produce sold, delivered or consigned by the grower from his holding for market preparation, packaging and storage;
(b) produce consigned from storage for market preparation and packaging;
(c) produce consigned from wholesale markets to preparation and packing stations or storage stations situated in the same production area. On secondary sale by retail other than sales on the producer's holding as below, produce must be properly graded and labelled;
(d) produce consigned to processing plants;
(e) produce transferred by the producer on his holding to consumers for their personal use.

1 See Appendix B.

Quality Standards

7.5.2 Each EC standard prescribes minimum marketing requirements and up to four quality classes – Extra Class, Class I, Class II and Class III. The quality classes may be summarised as:

Extra Class – excellent quality and usually only very specially selected and presented produce.
Class I – good quality produce with no important defects.
Class II – reasonably good quality, sound but deficient in one or two requirements such as shape, colour, smell, blemishes and marks.
Class III – produce of marketable quality which is not up to the standard of the higher classes. This class only operates in special circumstances such as shortage of supply throughout the Community. In recent years it has only been introduced for a limited period for witloof chicory and asparagus.

Labelling

7.5.3 Each package must bear the required particulars under the heading set out below in letters grouped on the same side, legibly and indelibly marked and visible from the outside:

(a) *Identification* – the name and address or officially issued or accepted code mark of the packer and/or dispatcher.
(b) *Nature of Produce* – if the contents are not visible from the outside (eg 'peas').
(c) *Origin of Produce* – the country of origin or, for a temporary period, the district of origin or national regional or local trade name.
(d) *Commercial Specifications* – quality class
– size and/or count

Chapter 8

Nutrition Labelling and Claims

8.1 Nutrition Labelling – EC Requirements

Nutrition Information

8.1.1 Where certain claims are made in the advertising or labelling of food the giving of certain prescribed nutrition information is compulsory.[1] In all other cases the giving of such information is voluntary. Conscious of the fact that voluntary nutrition information on labels was being given in a variety of different ways, using different terms and values, the Ministry of Agriculture, Fisheries and Food produced guidelines in July 1987 recommending the format and criteria for such information. The guidelines were revised in January 1988 and although they had no force in law they were widely observed. In September 1990 a Directive,[1] (the Directive) on nutrition labelling for foodstuffs was adopted. Member states are required to permit trade in products complying with the Directive from 1 April 1992 and to prohibit trade in products which do not comply from 1 October 1993.

The Directive requires that when energy-related claims are made for a food certain nutrition information shall be given in a prescribed form. When nutrition information is given voluntarily it too must be given in the prescribed form.

Until the Directive is implemented in the United Kingdom by law, the Ministry has recommended that nutrition information should be given as required by it and the guidelines issued in January 1988 should be disregarded. However, there are anomalies between the Directive and the claims requirements of the Food Labelling Regulations 1984 (see **8.2**). The provisions as to vitamins, for example, are to be reviewed with the probability that a new Annex to the Directive with different recommended daily amounts will be substituted from late 1992. Other matters of detailed interpretation have yet to be settled.

1 Council Directive 90/496/EEC (OJ No L 276; 06.10.90).

Foodstuffs for Particular Nutrition Purposes[1]

8.1.2 In May 1989 the Council of the EC adopted a Directive empowering the Commission to adopt vertical directives on the labelling and marketing of food for particular nutritional uses (PARNUTS). The Directive establishes a framework for a common system for governing PARNUTS foods throughout the Community. It introduces a notification procedure for new products and provides for the suspension or restriction of sales of foods claiming to be PARNUTS products but which do not meet the criteria laid down in the Directive. The Directive also introduces nine categories of food for which vertical Directives will be introduced by the EC Commission. United Kingdom regulations for a notification system by manufacturers and importers to their home food authority are imminent.

1 Council Directive 89/398/EEC (OJ No L 186; 30.06.89).

Compulsory Nutrition Labelling

8.1.3 The Directive requires that where a nutrition claim appears in labelling, in presentation or in advertising, nutrition labelling is compulsory. A nutrition claim is any representation or any advertising message which states, suggests or implies that a foodstuff has particular nutritional properties due to the energy (calorific value) it provides, or provides at a reduced or increased rate, or does not provide, and/or due to the nutrients it contains, or contains in reduced or increased proportions, or does not contain. A reference to qualities or quantities of a nutrient does not constitute a nutrition claim in so far as it is required by legislation.

8.1.4 *Nutrition labelling* is defined by the Directive as being any information appearing on labelling relating to:

(i) energy value
(ii) the following nutrients:
  - protein,
  - carbohydrate,
  - fat,
  - fibre,
  - sodium,
  - vitamins and minerals listed below and present in the significant amounts indicated.

Vitamins and Minerals which may be Declared and their Recommended Daily Allowances (RDAs)

Vitamin A μg	800	Vitamin B12 μg	1
Vitamin D μg	5	Biotin mg	0.15
Vitamin E mg	10	Pantothenic acid mg	6
Vitamin C mg	60	Calcium mg	800
Thiamin mg	1.4	Phosphorus mg	800
Riboflavin mg	1.6	Iron mg	14
Niacin mg	18	Magnesium mg	300
Vitamin B6 mg	2	Zinc mg	15
Folacin μg	200	Iodine μg	150

As a rule 15% of the RDA supplied by 100g or 100 ml, or per package if the package contains only a single portion, should be taken into consideration in deciding what constitutes a significant amount.

Form of Nutrition Labelling

8.1.5 Where nutrition labelling is provided the information must be given in either group 1 or group 2 as follows:

Group 1
(a) energy value
(b) the amounts of protein, carbohydrate and fat.

Group 2
(a) energy value
(b) the amounts of protein, carbohydrate, sugars, fat, saturates, fibre and sodium.

Where a nutrition claim is made for sugars, saturates, fibre or sodium, the information given must consist of group 2.

The nutrition labelling may also include the amounts of one or more of the following:

- starch,
- polyols,
- monounsaturates,
- polyunsaturates,
- cholesterol,
- any of the minerals or vitamins listed above and present in significant amounts.

Where sugars and/or polyols and/or starch are declared, this declaration shall immediately follow the declaration of the carbohydrate content in the following manner:

– carbohydrate	g
of which:	
– sugars	g
– polyols	g
– starch	g

Where the amount and/or type of fatty acid and/or the cholesterol rate is declared, this declaration shall immediately follow the declaration of total fats in the following manner:

– fat	g
of which:	
– saturates	g
– monounsaturates	g
– polyunsaturates	g
– cholesterol	mg

The declaration of substances which belong to or are components of one of the categories of nutrients referred to above are compulsory when a nutrition claim is made.

In addition, where the amount of polyunsaturates and/or monounsaturates and/or the cholesterol rate is given, the amount of saturates shall also be given, the declaration of the latter not constituting, in this case, a nutrition claim as defined above.

Energy value

8.1.6 The energy value to be declared must be calculated using the following conversion factors:

– carbohydrate (except polyols)	4 kcal/g – 17 kJ/g
– polyols	2.4 kcal/g – 10 kJ/g
– protein	4 kcal/g – 17 kJ/g
– fat	9 kcal/g – 37 kJ/g
– Alcohol (ethanol)	7 kcal/g – 29 kJ/g
– organic acid	3 kcal/g – 13 kJ/g

Units to be Used

8.1.7 The units to be used in declarations are:

– energy- kJ and kcal	
– protein	grams (g)
– carbohydrate	grams (g)
– fat	grams (g)
– fibre	grams (g)
– sodium	grams (g)
– cholesterol	milligrams (mg)
– vitamins and minerals	the units specified at 7.1.4 above.*

* The information on vitamins and minerals must also be expressed as a percentage of the RDA for each 100g or 100 ml.

All nutritional information must be expressed per 100g or 100 ml. In addition, this information may be given per serving as quantified on the label or per portion, provided that the number of portions contained in the package is stated.

Presentation of Information

8.1.8 The information covered by the Directive must be presented together in one place in tabular form, with the numbers aligned if space permits. Where space does not permit, the information shall be presented in linear form. It shall be printed in legible and indelible characters in a conspicuous place. The language used must be that which would be understood in the market concerned. Subject to this requirement, it may also appear in more than one language.

Declared Values

8.1.9 The declared values shall, according to the individual case, be average values based on:

(a) the manufacturer's analysis of the food;
(b) a calculation from the known or actual values of the ingredients used;
(c) a calculation from generally established and accepted data.

Accuracy of Nutritional Information

8.1.10 If nutrition information is misleading, whether it is given voluntarily or is required as a result of a claim having been made, an offence against the Food Safety Act 1990 (false or misleading label) or against the Trade Descriptions Act 1968 (false trade description) (see Chapter 6) may have been committed. Bearing in mind the criteria for assessment of values (**8.1.9**) the inaccuracy would have to be significant. There have been successful prosecutions on this matter.

8.2 Claims Controlled by the Food Labelling Regulations

Prohibited Claims[1]

8.2.1 Any claim, whether made expressly or by implication, that a food has tonic properties (the use of the word 'tonic' to denote a soft drink excepted) is prohibited. So too is a claim that a food which is intended for babies is equivalent or superior to the milk of a healthy mother.

1 Food Labelling Regulations 1984, reg 36 (1) and Part I of Sch 6.

Restricted Claims[1]

8.2.2 A claim of a type indicated below shall not be made, either expressly or by implication, in the labelling or advertising of a food except in accordance with the stated appropriate conditions.

1. *Claims relating to foods for particular nutritional uses*
A claim that a food is suitable, or has been specially made for fulfilling the particular nutritional requirements of –

(a) a class of persons whose digestive process or metabolism is disturbed, or
(b) a class of persons who, by reason of their special physiological condition, obtain special benefit from a controlled consumption of certain substances.

Conditions
1 The food must be capable of fulfilling the claim.
2 The food must be marked or labelled with an indication of the particular aspects of its composition or manufacturing process that give the food its particular nutritional characteristics.
3 If the food has been specially made for the class of persons to which the claim refers –
  (a) the name of the food must include or be accompanied by an indication of the food's particular nutritional characteristics,
  (b) the food must be marked or labelled with the prescribed energy statement*, and
  (c) when sold to the ultimate consumer, the food must be prepacked and completely enclosed by its packaging.

2. *Claims relating to babies or young children***
A claim that a food is suitable, or has been specially made, for fulfilling the particular nutritional requirements of babies or young children.

Conditions
1 The food must be capable of fulfilling the claim.
2 If the food has been specially made for babies or young children –
  (a) the food must be marked or labelled with an indication that it is intended for babies or young children,
  (b) the food must be marked or labelled with the prescribed energy statement,*
  (c) when sold to the ultimate consumer the food must be prepacked and completely enclosed by its packaging.

3. *Diabetic claims*
A claim that a food is suitable, or has been specially made, for diabetics.

Conditions

1(1) A given quantity of the food must not have a higher energy content than the same quantity of a similar food in relation to which no diabetic claim is made, unless the food is –
  (a) an intense sweetener,
  (b) a product which consists of a mixture of an intense sweetener with other substances and which, when compared on a weight for weight basis, is significantly sweeter than sucrose.

(2) A given quantity of the food must not have a higher fat content than the same quantity of a similar food in relation to which no diabetic claim is made.

(3) In the case of a claim that a food has been specially made for diabetics, a given quantity of the food must not have a readily absorbable carbohydrate content greater than 50% of the readily absorbable carbohydrate content of the same quantity of a similar food in relation to which no diabetic claim is made.

(4) The food must not contain a greater quantity of mono- or disaccharides, other than fructose, than the quantity which is technically necessary to retain the essential characteristics of the food whilst having regard to its claimed suitability for diabetics.

Where a food is in concentrated or dehydrated form and is intended to be reconstituted by the addition of water or other substances, this condition applies to the food when reconstituted as directed.

2(1) If the claim is made in the labelling of the food, the food must be marked or labelled with a statement to the effect that the food is not suitable for an overweight diabetic, unless the energy value of a given weight of the food is not more than one half of that of the equivalent weight of a similar food in relation to which no diabetic claim is made.

(2) If the claim is made in the advertising of a food, otherwise than in an oral advertisement (including a television advertisement), either –
  (a) the advertisement must, subject to the same qualification, include the said statement, or
  (b) subject to the same qualification, the food in relation to which the claim is made must be intended to be sold prepacked and, when sold, must be marked or labelled with the said statement.

(3) If the claim has been made in the labelling or advertising of a food that contains fructose, sorbitol, mannitol, xylitol, isomalt, lactitol or hydrogentated glucose syrup, the food must be marked or labelled with the words –
  (a) 'best eat less than 25 g of X a day', where only one such substance is contained in the food, there being inserted in place of 'X' the name of that substance,
  (b) 'best eat less than 25 g of a combination of X a day', where more than one such substance is contained in the food, there being inserted in place of 'X' the name of every such substance that the food contains.

3 The conditions set out in item 1 above must be satisfied, except that –
  (a) condition 3(b) of item 1 must be satisfied whether or not the food has been specially made for diabetics, and
  (b) if the food contains fructose, sorbitol, mannitol, xylitol, isomalt, lactitol or hydrogenated glucose syrup, the prescribed energy statement* must include a statement of that total amount of –
    (i) the aforementioned substances that are present in the food, and

(ii) other carbohydrate, and
(iii) the total amount of carbohydrate that is contained in each 100g or 100 ml of the food, as is appropriate, and, where appropriate, in a quantified serving of the food.

Note: 'Readily absorbable carbohydrate' does not include sugar alcohols or fructose, but includes other mono- and disaccharides and isolated or purified starches.

4. *Slimming claims*
A claim that a food is an aid to slimming or weight control or weight reduction or has a reduced or low energy value (the words 'low calorie' on a container of soft drink in accordance with reg 11 of the Soft Drinks Regulations 1964 is not of itself a claim).

Conditions

1 The food must be capable of contributing to weight control or weight reduction.

2 The food must not consist wholly or mainly of vitamins or minerals or of a mixture of vitamins and minerals.

3 The conditions set out in item 1 above must be satisfied, except that in the case of prepacked food condition 3(b) of item 1 must be satisfied whether or not the food has been specially made for such a class of persons as specified in item 1.

4(1) If the claim is made in the labelling of the food, the food must be marked or labelled with the words 'can help slimming or weight control only as part of a controlled diet', except that the word 'calorie', 'joule' or 'energy' may be inserted immediately before the word 'controlled'.

(2) If the claim is made in the advertising of the food, the advertisement must include the said words, subject to the same qualification.

5 If the claim is that a food has a reduced energy value, the energy value of a given weight of the food, or a given volume in the case of a liquid food, must not be more than three quarters of that of the equivalent weight or volume of a similar food in relation to which no such claim is made, unless the food is –
(a) an intense sweetener,
(b) a product which consists of a mixture of an intense sweetener with other substances and which, when compared on a weight for weight basis, is significantly sweeter than sucrose.

6 If the claim is that the food has a low energy value –
(a) the energy value of the food must not be more than 167 kJ (40 kcal) per 100g or 100 ml as appropriate, unless the food is –
(i) an intense sweetener, or
(ii) a product which consists of a mixture of an intense sweetener with other substances and which, when compared on a weight for weight basis, is significantly sweeter than sucrose,
(b) the energy value of a normal serving of the food must not be more than 167 kJ (40 kcal), and
(c) in the case of an uncooked food which naturally has a low energy value, the claim must be in the form 'a low energy food' or 'a low calorie food' or 'a low joule food'.

7 If the claim is combined with a claim that the food is suitable to replace part of a daily diet, the food must be marked or labelled with a statement to the effect that the food –

(a) should not be used as a replacement for a person's whole diet, and
(b) should be supplemented by at least one daily meal having a high nutritive quality.

Note: Where a food is in concentrated or dehydrated form and is intended to be reconstituted by the addition of water or other substances, conditions 6 and 7 apply to the food when reconstituted as directed.

5. *Medicinal claims*
A claim that a food is capable of preventing, treating or curing human disease (this does not include a claim of a type in items 1 or 2 above).

Conditions
The food must have a product licence issued under the provisions of the Medicines Act 1968.

6. *Protein claims*
A claim that a food, other than a food intended for babies or young children** which satisfies the conditions of item 1 above, is a source of protein.

Conditions
1 The quantity of the food that can reasonably be expected to be consumed in one day must contribute at least 12 g of protein.
2(1) If the claim is that the food is a rich or excellent source of protein, at least 20% of the energy value of the food must be provided by protein.
(2) In any other case at least 12% of the energy value of the food must be provided by protein.
3(1) If the claim is made in the labelling of the food, the food must be marked or labelled with a statement of the protein content of each hundred 100 g or 100 ml of the food, as appropriate, and, where appropriate, of a quantified serving of the food.
(2) If the claim is made in the advertising of the food, the food when sold prepacked must be marked or labelled with the said statement.

7. *Vitamin claims*
A claim that a food other than a food intended for babies or young children** which satisfies the conditions of item 1 above, is a source of vitamins.

Note: A reference to a vitamin in the name of a food shall not of itself constitute a claim of a type to which this item applies if the food consists solely of –

(i) vitamins, or
(ii) a mixture of vitamins and minerals, or
(iii) a mixture of vitamins, or vitamins and minerals, and a carrying agent, or
(iv) a mixture of vitamins, or of vitamins and minerals, and other substances sold in tablet, capsule or elixir form.

Conditions
1(1) If the claim is not confined to named vitamins, every vitamin named in the claim must be a vitamin specified in column 1 of Table A below, and –
(a) where the claim is that the food is a rich or excellent source of vitamins, the quantity of the food that can reasonably be expected to be consumed in one day must contain at least one half of the RDA of two or more of the vitamins specified in column 1 of Table A below, and

(b) in any other case, the quantity of the food that can reasonably be expected to be consumed in one day must contain at least one sixth of the RDA of two or more of the vitamins specified in column 1 of Table A below.

(2) If the claim is confined to named vitamins, every vitamin named in the claim must be a vitamin specified in column 1 of Table A below, and –

(a) where the claim is that the food is a rich or excellent source of vitamins, the quantity of the food that can reasonably be expected to be consumed in one day must contain at least one half of the RDA of every vitamin named in the claim, and

(b) in any other case the quantity of the food that can reasonably be expected to be consumed in one day must contain at least one sixth of the RDA of every vitamin named in the claim.

2(1) If the claim is made in the labelling of a food, the food must be marked or labelled with a statement–

(a) (i) where the claim is not confined to named vitamins, of the percentage of the RDA of the vitamins enabling the claim to be made that is contained in a quantified serving of the food; and

(ii) where the claim is confined to named vitamins, of the percentage of the RDA of every vitamin named in the claim that is contained in a quantified serving of the food, and

(b) where the food is prepacked, of the number of such servings contained in the package, and the name used in the statement for any such vitamin shall be the name specified for that vitamin in column 1 of Table A below, with or without the words, if any, that appear in brackets in that column of that Table.

(2) If the claim is made in the advertising of a food, the food when sold prepacked must be marked or labelled with the said statement.

8. *Mineral claims*

A claim that a food, other than a food intended for babies or young children which satisfies the conditions of item 1 above, is a source of minerals. A claim that a food has a low or reduced levels of minerals shall not be regarded as a claim of a type described in this item. The note to item 8 above applies equally to this item with the substitution of the word 'mineral' for 'vitamin' and vice versa as appropriate.

Conditions

The conditions are the same as those set out in item 7 above with appropriate substitution of the word 'mineral' for 'vitamin' and 'Table B' for 'Table A'.

9. *Polyunsaturated fatty acid claims*

A claim relating to polyunsaturated fatty acids in a food.

Conditions

1(a) The food must contain at least 35% fat by weight, and

(b) at least 45% of the fatty acids must be polyunsaturated and not more than 25% of the fatty acids may be saturated; and

(c) the claim must be accompanied by the words 'low in saturates' or 'low in saturated fatty acids'; and

(d) the food must be marked or labelled with a declaration, expressed in g per 100 g or ml per 100 ml of the food as is appropriate, stating –

(i) the amount of fat or oil;

(ii) the amount of polyunsaturated fatty acids which are *cis*, *cis*-methylene interrupted polyunsaturated fatty acids; and

(iii) the amount of saturated fatty acids; and each part of the declaration must be given equal prominence.

2 The claim must not be accompanied by a suggestion, whether express or implied, that the food is beneficial to human health because it contains polyunsaturated fatty acids.

10. *Cholesterol claims*

A claim relating to the presence or absence of cholesterol in a food.

Conditions

1 The food must contain no more than 0.005% of cholesterol.
2 A polyunsaturated acid claim must be made and the conditions set out in item 9 above satisfied.
3 The claim relating to cholesterol must not be in larger letters than, nor precede, the polyunsaturated fatty acid claim.
4 The claim must not be accompanied by a suggestion, whether express or implied, that the food is beneficial to human health because of its level of cholesterol.

11. *Energy claims*

A claim that a food provides more energy than other food or is a good source of energy.

Conditions

1 The food must be capable of fulfilling the claim.
2 If the claim is made in the labelling of any food, the food must be marked or labelled with the prescribed energy statement*.
3 If the claim is made in the advertising of the food, the food, when sold prepacked, must be marked or labelled with the prescribed energy statement*.

12. *Claims which depend on another food*

The value or benefit must not be derived wholly or partly from another food that is intended to be consumed with the food in relation to which the claim is made.

* Under the Food Labelling Regulations 1984, reg 37, the prescribed energy statement means –

(a) In relation to a food with an energy value of 50 kilojoules (12 kilocalories) or more per 100 g or 100 ml, as the case may be, a statement of –
   (i) the energy value expressed in kilojoules and kilocalories, of each 100 g or 100 ml of the food, as appropriate, and where appropriate, of a quantified serving of the food, and
   (ii) the amount of carbohydrate, protein and fat contained in each 100 g or 100 ml of the food as appropriate, and, where appropriate, in a quantified serving of the food, and

(b) in relation to a food with an energy value of less than 50 kilojoules (12 kilocalories) per 100 g or 100 ml, as the case may be, either –
   (i) a statement of the kind described in sub-paragraph (a) above, or
   (ii) a statement that the energy value of the food is less than 50 kilojoules (12 kilocalories) per 100 g or 100 ml, as is appropriate.

In the calculation of the energy value of a food the following conversion factors should be employed:

(a) 1 gram of carbohydrate (expressed as monosaccharides) shall be deemed to contribute 16 kilojoules (3.75 kilocalories);
(b) 1 gram of protein shall be deemed to contribute 17 kilojoules (4 kilocalories);

(c) 1 gram of alcohol shall be deemed to contrbute 29 kilojoules (7 kilocalories);
(d) 1 gram of fat shall be deemed to contribute 37 kilojoules (9 kilocalories).

** The term 'babies or young children' is not defined. A young child would appear to be one whose digestive processes have not yet reached maturity.

Nothing in the provisions as to claims should be taken to prevent the dissemination of useful information or recommendations intended exclusively for persons having qualifications in dentistry, medicine, nutrition, dietetics or pharmacy.

The following do not of themselves constitute a claim of a type discussed above:

(a) a reference to a substance other than vitamins, minerals, polyunsaturated fatty acids or cholesterol in the name of a food;
(b) a reference to a substance in a list of ingredients or in a statement of the total nutrient content of a food;
(c) a statement of the energy value of a food.

1 Food Labelling Regulations 1984, reg 35 (2) and Part II of Sch 6.

Vitamins and Minerals

8.2.3 The Tables referred to in paragraph **8.2.2** above in relation to vitamins and minerals are as follows:

Table A

Vitamin	*To be calculated as*	*Recommended Daily amount*
Vitamin A	Micrograms of retinol or micrograms of retinol equivalent on the basis the 6 μg of beta-carotine or 12 μg of other biologically active carotenoids equal 1 μg of retinol equivalent	750 μg
Thiamin (vitamin B1)	Milligrams of thiamin	1.2 mg
Riboflavin (vitamin B2)	Milligrams of riboflavin	1.6 mg
Niacin	Milligrams of nicotinic acid or milligrams of nicotin-amide or millgrams of niacin equivalent on the basis that 60 mg of tryptophan equal 1 mg of niacin equivalent	18 mg
Folic acid	Micrograms of total folic acid	300 μg
Vitamin B12	Micrograms of cobalamines	2 μg
Vitamin C (ascorbic acid)	Milligrams of ascorbic acid or milligrams of dehydroascorbic acid.	30 mg
Vitamin D	Micrograms of ergocalciferol (vitamin D2) or micrograms of cholecalciferol (vitamin D3)	2.5 μg

Table B

Mineral	*To be calculated as*	*Recommended daily amount*
Calcium	Milligrams of calcium	500 mg
Iodine	Micrograms of iodine	140 μg
Iron	Milligrams of iron	12 mg

Notes
1 Each vitamin and mineral specified above includes its biologically active derivative.
2 The quantity of any vitamin or mineral specified above (as extended by note 1 above) shall be calculated in accordance with column 2 of the appropriate Table.

'Natural Claims'

8.2.4 The use of the word 'natural' and similar terms in food labelling and advertising has been a source of concern for some years. In July 1989 the Ministry of Agriculture, Fisheries and Food, having first consulted the Food Advisory Committee, issued guidelines on such claims. It was made clear that if the guidelines were not successful in curbing the worst excesses in such matters, regulations may follow.

The guidelines suggested six principles on which such labelling should be based:

1 The term 'natural' without qualification should be used only:
 (a) to describe single foods, of a traditional nature to which nothing has been added and which have been subjected to such processing as to render them suitable for human consumption. Freezing, concentration, fermentation, pasteurisation, sterilisation, smoking (without chemicals) and traditional cooking processes such as baking, roasting and blanching are processes which would be acceptable within this criteria. Bleaching, oxidation, smoking with chemicals, hydrogenation and similar processes are outside the criteria. As a general rule for single ingredient foods such as cheese, yogurt and butter, acceptable processing would be that which is strictly necessary to produce the final product.
 The restriction to food 'of a traditional nature' is intended to exclude foods such as mycoprotein, which may technically be products of natural sources but which are considered not to accord with the public perception of 'natural';
 (b) to describe food products obtained from recognised food sources and which meet the criteria in (a) above;
 (c) to describe flavouring substances or permitted food additives obtained from recognised food sources by appropriate physical processing (including distillation and solvent extraction) or traditional food preparation processes. Flavours should only be so described when they are derived wholly from the named food source.
2 Compound foods should not therefore be described directly or by implication as 'natural', but it may be acceptable to describe such foods as 'made from natural ingredients' if all the ingredients meet the criteria in 1(b) or (c).
3 A food which does not meet the criteria in 1(a) or 2 above should not be claimed to have a 'natural' taste, flavour or colour.
4 'Natural' or its derivatives, should not be included in brand or fancy names nor in coined or meaningless phrases in such a way as to imply that a food which does not meet the criteria in 1(a) is natural or made from natural ingredients.

5 Claims such as 'natural goodness', 'naturally better' or 'nature's way' are largely meaningless and should not be used.
6 'Natural' meaning no more than plain or unflavoured should not be used except where the food in question meets the criteria in 1(a) or 2.

The principles above apply equally to other words such as 'real', 'genuine' and 'pure' which although having separate and distinctive meanings of their own should not be used in place of 'natural' in such a way as to imply similar benefits for consumers. In the latter circumstances the principles set out above should apply to such words.

In addition there are a variety of claims (which might be termed 'negative claims') which do not use the term 'natural' or its derivatives, but the effect of which is to imply 'naturalness' to the consumer. Such claims are potentially misleading and confusing to the consumer and at least the following should not be used:

(a) a claim that a food is 'free from X' if all foods in the same class or category are free from 'X';
(b) statements or implications which give undue emphasis to the fact that a product is free from certain non-natural additives when the product contains other non-natural additives;
(c) a claim that a food is 'free from' one category of additive when an additive of another category, or an ingredient having broadly similar effect is used.

However, it is recognised that certain 'negative claims' which do not imply 'naturalness' to the consumer, such as 'free from "X"', where 'X' is a particular additive, may provide accurate and beneficial information for consumers.

Product surveys carried out by trading standards officers during 1991 suggest that these guidelines have been widely disregarded.

Organic Claims

8.2.5 In June 1991 a regulation[1] on organic indications in relation to agricultural products and foodstuffs was adopted and must be given the force of law in its entirety in the United Kingdom by 23 July 1992.

Claims that a food has been produced organically may only be made where prescribed principles of organic production at farm level have been satisfied. The principles are:

1. The fertility and biological activity of the soil must be maintained or increased, where appropriate, by:
   (a) cultivation of legumes, green manures or deep rooting plants in an appropriate multiannual rotation programme;
   (b) incorportion in the soil of organic material, composted or not, from holdings producing according to the rules of the regulation. Pending the adoption of common technical rules concerning organic livestock production, by-products from livestock farming, such as farmyard manure, may be used if they come from livestock holdings respecting existing national rules or, in the absence thereof, internationally recognised practices concerning organic livestock production.

   Other organic or mineral fertilisers (listed in the Regulation) may be applied only to the extent that adequate nutrition of the crop being rotated or soil conditioning is not possible by the methods set out under (a) and (b) above.

For compost activation, appropriate micro-organism or plant based preparations (biodynamic preparations) may be used.

2. Pests, diseases and weeds shall be controlled by a combination of the following measures:
   - choice of appropriate species and varieties,
   - appropriate rotation programmes,
   - mechanical cultivation procedures,
   - protection of natural enemies of pests through provisions favourable to them (eg hedges, nesting sites, release of predators),
   - flame weeding.

   In cases of immediate threat to the crop recourse may be had to certain natural and mineral materials listed in the regulation.

National organic food standards have been produced in the United Kingdom by UKROFS.[2]

1 Council Regulation (EEC) No 2092/91 (OJ No L 198; 22.07.91).

2 United Kingdom Register of Organic Food Standards, 301–344 Market Towers, New Covent Garden Market, London SW8.

Other Claims

8.2.6 For the past 10 years the most influential guide to general claims in food labelling and advertising has been the Second Report of the Food Standards Committee (now the Food Advisory Committee) on Claims and Misleading Descriptions.[1] Some of the recommendations in that report were adopted in the Food Labelling Regulations 1984 and others have been the subject of subsequent advice and guidance notes. However, in its most recent report[2] the Committee has established a series of criteria as to food claims of which the following is a brief summary:

1. A food must be able to fulfil the claim being made for it and adequate labelling information should be given to show consumers that the claim is justified.
2. Where the claim is potentially ambiguous or imprecise (for example 'light') it should be clearly explained and justified on the label and be capable of substantiation.
3. A claim that a food is 'free from' a substance or treatment should not be made if all the same class or category of foods are similarly free.
4. Words or phrases which imply that a food is free from any specific characteristic ingredient or substance should not be used if the food contains other ingredients or substances with the same characteristic.
5. Meaningless descriptions should not be used.
6. Comparative claims should be justified against relative and generally acceptable criteria.
7. The label should give a sufficiently full description of the food in relation to the area for which the claim is made to ensure that selective claims, even if true, do not mislead.
8. Absolute claims should be justified against absolute criteria set for a given nutrient and applying to all foods.
9. Where a food is naturally 'low' or 'high' in a substance the claim should be 'a low/high X food'.

8.2.6 *Nutrition Labelling and Claims*

The Committee also considered medicinal and health claims and recommended that the present law[3] should be maintained and claims relating to a deficiency disease, which is virtually non-existent in the EC, should be prohibited.

Health claims should be made in accordance with the following general principles:

1. The claim should relate to the food as eaten rather than to generic properties of any of the ingredients.
2. The food when consumed in normal dietary quantities should be able to fulfil the claim being made for it and adequate labelling information should be given to show consumers that the claim is justified.
3. The label should give a full description of the food to ensure that selective claims, even if true, do not mislead and any claim should trigger full nutritional labelling (see **8.1**) above.
4. The role of the specific food should be explained in relation to the overall diet and other factors.

Tentative EC proposals for a Directive on Food Claims were produced in 1988 and it may be expected that further proposals will emerge during 1992/3. Meanwhile, with public and enforcement attention focussed on food claims, manufacturers should be cautious about the claims they make and have regard to the FAC recommendations above.

1 FSC/REP 71 (MAFF, 1980).
2 Food Advisory Committee Report on its Review of Food Labelling and Advertising (1990).
3 Food Labelling Regulations 1984, Sch 6 and the Medicines Act 1968, Part VI.

Chapter 9

Additives, Flavourings, Preparation and Contaminants

9.1 Additives

Current Controls

9.1.1 Current controls on food additives are all based on the positive list principle; that is, no additives other than those on a list of permitted additives may be used in or on food for human consumption. For legislative purposes additives are divided into categories and current United Kingdom legislation (see Appendix A) implements EC requirements (see Appendix B). The categories of additives so controlled are antioxidants, colours, emulsifiers and stabilisers, erucic acid, extraction solvents, preservatives and miscellaneous additives, mineral hydrocarbons and sweeteners. Additives which are permitted for use on 1 January 1992 are listed in Appendix C and the labelling requirements for additives are at **6.4.13**.

Maximum Amounts

9.1.2 Maximum amounts of permitted additives which may be used in specified foods are prescribed in many cases. Where no such maximum is stated in the regulations the amounts used must accord with good manufacturing practice. In practice, the excessive use of an additive which is not subject to any maxima in the regulations could give rise to an offence against the Food Safety Act 1990.[1]

1 Food Safety Act 1990, ss 8 or 14.

EC Framework Directives

9.1.3 EC framework directives for additives[1] and flavourings[2] were adopted by the Council in February 1989 and July 1988 respectively. Since those dates the Commission has been working on new vertical directives for each category of additives and adoption of the first of those directives is expected towards the end of 1992. The positive list principle will be retained with variations and, in the case of flavourings, negative lists, that is flavourings which may not be used either generally or for particular foods, may be employed.

The first directive to be made will cover all food additives other than colours and sweeteners. These latter categories will be dealt with by individual directives at a later stage and new categories will be introduced (see **9.1.5**). It is anticipated that most of the additives permitted under existing legislation and listed in Appendix C will continue to be permitted for use in food but there will be new restrictions on the food in which certain additives may be used. This is especially the case for food colours. Another innovation will be new general controls on the use of additives in infant formula, follow-on formula and weaning foods. A system of conditionally permitted additives will also be introduced. Serial numbers allocated to additives will not be changed but, of course, new 'E' numbers will be allocated to additives from time to time.

When the new verticial directives are in place all existing United Kingdom additives regulations will be revoked and replaced by new regulations implementing the relevant provisions of the framework directives and the new lists of permitted additives. This is likely to occur by the end of 1993. The flavourings Directive, as amended, contains provisions as to the labelling and description of flavourings intended for sale to the ultimate consumer which must be implemented in the United Kingdom by 30 June 1992. Implementing regulations are imminent.

1 Council Directive 89/107/EEC (OJ No L 040; 11.02.89).
2 Council Directive 88/388/EEC (OJ No L 184; 15.07.88).

Definitions

9.1.4 New definitions for additives and flavourings have been introduced. They are:

- 'Food additive' means any substance not normally consumed as a food in itself and not normally used as a characteristic ingredient of food whether or not it has nutritive value, the intentional addition of which to food for a technological purpose in the manufacture, processing, preparation, treatment, packaging, transport or storage of such food results, or may be reasonably expected to result, in that substance or its by-products becoming directly or indirectly a component of such foods.
- 'Flavouring' means flavouring substances, flavouring preparations, process flavourings, smoke flavourings or mixtures thereof.
- 'Flavouring substance' means a defined chemical substance with flavouring properties which is obtained:
  - (i) by appropriate physical process (including distillation and solvent extraction) or enzymatic or microbiological processes from material of vegetable or animal origin either in the raw state or after processing for human consumption by traditional food preparation processes (including drying, torrefaction and fermentation);
  - (ii) by chemical synthesis or isolated by chemical processes and which is chemically identical to a substance naturally present in material of vegetable or animal origin as described in (i) above;
  - (iii) by chemical synthesis but which is not chemically identical to a substance naturally present in material of vegetable or animal origin as described in (i) above.
- 'Flavouring preparation' means a product, other than the substances defined in sub-paragraph (i) of the definition of 'flavouring substance', whether concentrated or not, with flavouring properties, which is obtained by appropriate physical processes (including distillation and solvent extraction) or by enzymatic or microbiological processes from material of animal or vegetable origin, either in the raw state or after processing for human consumption by traditional food preparation processes (including drying, torrefaction and fermentation).

- 'Process flavouring' means a product which is obtained according to good manufacturing processes by heating to a temperature not exceeding 180°C for a period not exceeding 15 minutes a mixture of ingredients, not necessarily themselves having flavouring properties, of which at least one contains nitrogen (amino) and another is a reducing sugar.
- 'Smoke flavouring' means a smoke extract used in traditional foodstuffs smoking processes.

Categories for Food Additives

9.1.5 Categories of food additives are laid down in the Directive. They are:

Colour
Preservative
Antioxidant
Emulsifier
Emulsifying salt
Thickener
Gelling Agent
Stabilizer (including foam stabilisers)
Flavour enhancer
Acid
Acidity regulator (can act as two-way acidity regulators)
Anti-caking agent
Modified starch
Sweetener
Raising agent
Anti-foaming agent
Glazing agent (including lubricants)
Flour treatment agent
Firming agent
Humectant
Sequestrant (without prejudice to any further decision or mention thereof in the labelling of foodstuffs)
Enzyme (without prejudice to any further decision or mention thereof in the labelling of foodstuffs; only those used as additives)
Bulking agent
Propellant gas and packaging gas.

Criteria for the Use of Food Additives

9.1.6 Criteria for the use of food additives are laid down in the Directive. These are the criteria to be used by national governments in consultation with the Commission when deciding whether a particular additive should be approved for use in food. The criteria are:

1. Food additives can be approved only provided that:
   - there can be demonstrated a reasonable technological need and the purpose cannot be achieved by other means which are economically and technologically practicable,
   - they present no hazard to the health of the consumer at the level of use proposed, so far as can be judged on the scientific evidence available;
   - they do not mislead the consumer.
2. The use of food additives may be considered only where there is evidence that the proposed use of the additive would have demonstrable advantages of benefit to the consumer. In other words, it is necessary to establish the case for what is commonly referred to as 'need'. The use of food additives should serve one or more of the purposes set out below and only where those purposes cannot be achieved by other means which are economically and technologically practicable and do not present a hazard to the health of the consumer:
   (a) to preserve the nutritional quality of the food; an intentional reduction in the nutritional quality of a food would be justified only where the food

does not constitute a significant item in a normal diet or where the additive is necessary for the production of foods for groups of consumers having special dietary needs;
   (b) to provide necessary ingredients or constituents for foods manufactured for groups of consumers having special dietary needs;
   (c) to enhance the keeping quality or stability of a food or to improve its organoleptic properties, provided that this does not so change the nature, substance or quality of the food as to deceive the consumer;
   (d) to provide aids in manufacture, processing, preparation, treatment, packing, transport or storage of food, provided that the additive is not used to disguise the effects of the use of faulty raw materials or of undesirable (including unhygienic) practices or techniques during the course of any of these activities.
3. To assess the possible harmful effects of a food additive or derivatives thereof, it must be subjected to appropriate toxicological testing and evaluation. The evaluation should also take into account, for example, any cumulative, synergistic or potentiating effect of its use and the phenomenon of human intolerance to substances foreign to the body.
4. All food additives must be kept under continuous observation and must be re-evaluated whenever necessary in the light of changing conditions of use and new scientific information.
5. Food additives must at all times comply with the approved criteria of purity.
6. Approved food additives must:
   (a) specify the foodstuffs to which these additives may be added and the conditions under which they may be added;
   (b) be limited to the lowest level of use necessary to achieve the desired effect;
   (c) take into account any acceptable daily intake, or equivalent assessment, established for the food additive and the probable daily intake of it from all sources. Where the food additive is to be used in foods eaten by special groups of consumers, account should be taken of the possible daily intake of the food additive by consumers in those groups.

Special Measures as to Flavourings

9.1.7 The flavourings Directive requires the EC Council to adopt appropriate provisions concerning:

- flavouring sources composed of foodstuffs, and of herbs and spices normally considered as foods;
- flavouring sources composed of vegetable or animal raw materials not normally considered as foods;
- flavouring substances obtained by appropriate physical processes or by enzymatic or microbiological processes from vegetable or animal raw materials;
- chemically synthesised or chemically isolated flavouring substances chemically identical to flavouring substances naturally present in foodstuffs or in herbs and spices normally considered as foods;
- chemically synthesised or chemically isolated flavouring substances chemically identical to flavouring substances naturally present in vegetable or animal raw materials not normally considered as foods;
- chemically synthesised or chemically isolated flavouring substances other than those referred to above;

- source materials used for the production of smoke flavourings or process flavourings, and the reaction conditions under which they are prepared;
- the use and methods of production of flavourings, including physical processes or enzymatic or microbiological processes for the production of flavouring preparations and flavouring substances for the protection of public health or trade;
- the conditions for the use of additives necessary for the storage and use of flavourings; products used for dissolving and diluting flavourings; and additives necessary for the production of flavourings (processing aids).

Labelling of Flavourings

9.1.8 Labelling of flavourings for sale to the final consumer must include the following information in visible, clearly legible and indelible form:

(a) either the word 'flavouring' or a more specific name or description of the flavouring;
(b) either the words 'for foodstuffs' or a more specific reference to the foodstuff for which the flavouring is intended;
(c) the date of minimum durability (in accordance with the Food Labelling Regulations 1984 – see **6.4.16**);
(d) where necessary, the special conditions for storage or use;
(e) instructions for use if omission thereof would prevent appropriate use of the flavouring;
(f) the net quantity expressed in units of mass or volume;
(g) the name or business name and address of the manufacturer or packer, or of a seller established within the EC;
(h) a lot or batch identification mark;
(i) in the case of a mixture of flavouring(s) with other substances a list in descending order by weight in the mixture of:
   - the flavouring(s) in question in accordance with (a) above;
   - the names of each of the other substances or materials or, where appropriate, their E numbers.

Natural Flavourings

9.1.9 The word 'natural' or any other word having substantially the same meaning, may be used only for flavourings in which the flavouring component contains exclusively flavouring substances and/or flavouring preparations. If the sales description of the flavourings contains a reference to a foodstuff or a flavouring source, the word 'natural' or any other word having substantially the same meaning, may not be used unless the flavouring component has been isolated by appropriate physical process, enzymatic or microbiological processes or traditional food-preparation processes solely or almost solely from the foodstuff or the flavouring source concerned.

The provisions of **9.1.8** and this paragraph will come into force by regulations in the United Kingdom from 30 July 1992.

Control of Certain Substances in Flavourings

9.1.10 Maximum limits for certain undesirable substances present in foodstuffs as consumed as a result of the use of flavourings are laid down in the Flavourings Directive.

The following maximum limits are laid down by Council Directive 88/388/EEC and will be implemented by regulations in the United Kingdom.

Maximum Limits for Certain Undesirable Substances Present in Foodstuffs as Consumed, as a Result of the Use of Flavourings

Substance	*Foodstuffs*	*Beverages*
3,4 benzopyrene	0.03 μg/kg	0.03 μg/kg

Maximum Limits for Certain Substances Obtained from Flavourings and other Food Ingredients with Flavouring Properties Present in Foodstuffs as Consumed in which Flavourings have been Used.

Substances	*Foodstuffs mg/kg*	*Beverages mg/kg*	*Exceptions and/or special restrictions*
Agaric acid*	20	20	100 mg/kg in alcoholic beverages and food containing mushrooms
Aloin*	0.1	0.1	50 mg/kg in alcoholic beverages
Beta asarone*	0.1	0.1	1 mg/kg in alcoholic beverages and seasonings in snack foods
Berberine*	0.1	0.1	10 mg/kg in alcoholic beverages
Coumarin*	2	2	10 mg/kg in certain caramel confectionery; 50 mg/kg in chewing gum; 10 mg/kg in alcoholic beverages
Hydrocyanic acid*	1	1	50 mg/kg in nougat marzipan or its substitutes or similar products; 1 mg/% vol of alcohol in alcoholic beverages
Hypericine*	0.1	0.1	10 mg/kg in alcoholic beverages; 1 mg/kg in confectionery
Pulegone*	25	100	250 mg/kg in mint or peppermint-flavoured beverages
Quassine*	5	5	10 mg/kg in confectionery in pastille form; 50 mg/kg in alcoholic beverages
Safrole and isosafrole*	1	1	2 mg/kg in alcoholic beverages with not more than 25% vol alc; 5 mg/kg in alcoholic beverages with more than 25% vol alc; 15 mg/kg in food containing mace and nutmeg
Santonin*	0.1	0.1	1 mg/kg in alcoholic beverages with more than 25% vol alcohol

Substances	*Foodstuffs mg/kg*	*Beverages mg/kg*	*Exceptions and/or special restrictions*
Thuyone (alpha and beta)	0.5	0.5	5 mg/kg in alcoholic beverages with more than 25% vol alc; 10 mg/kg in alcoholic beverages with with more than 25% vol alc; 25 mg/kg in food containing preparations based on sage; 35 mg/kg in bitters

* May not be added as such to foodstuffs or to flavourings. May be present in a foodstuff either naturally or following the addition of flavourings prepared from natural raw material.

9.2 Preparation

9.2.1 The preparation, processing or treatment of any food is now firmly controlled by the Food Safety Act 1990. Ministers now enjoy extensive powers to control or prohibit undesirable or potentially dangerous food preparation techniques. There are general offences applicable to such preparation of food and certain forms of preparation are subject to regulations.

The term 'preparation' is defined[1] as including manufacture and any form of processing or treatment. Treatment is further defined as including the subjecting of food to heat or cold. The preparation of food is firmly locked into the overall control system by its inclusion in the key definition of a 'commercial operation'[2] and the important presumption as to 'food for human consumption'.[3]

1 Food Safety Act 1990, s 53 (1).
2 Food Safety Act 1990, s 1 (3).
3 Food Safety Act 1990, s 3 (3).

Criminal Liability

9.2.2 Criminal liability in respect of the preparation of food under the Food Safety Act 1990 includes the subjecting of any food to a process or treatment so as to render it injurious to health;[1] breach of the food safety requirement;[2] failing to comply with an improvement notice applicable to unlawful processing, treatment or preparation of food;[3] breach of a prohibition order as to any process or treatment;[4] and breach of an emergency prohibition notice and/or order.[5] Whether or not a person has prepared food is a crucial matter in determining the degree of proof required to establish a 'due diligence' defence.[6]

1 Food Safety Act 1990, s 7.
2 Food Safety Act 1990, s 8.
3 Food Safety Act 1990, s 10.
4 Food Safety Act 1990, s 11.
5 Food Safety Act 1990, s 12.
6 Food Safety Act 1990, s 21.

Ice-cream

9.2.3 The ingredients used in the manufacture of ice-cream must be pasteurised by one or other of three specified methods or sterilised and thereafter be kept at a specified low temperature until the freezing process is begun. It is an offence to sell or offer for sale ice-cream which has not been so treated or which has been allowed to reach a temperature exceeding 28°F without again being treated. Exemptions are provided from pasteurisation or sterilisation for certain types of water ices and ice lollies which are sufficiently acid to make such treatment unnecessary.[1]

1 Ice-cream (Heat Treatment etc) Regulations 1959/734, as amended.

Liquid Egg

9.2.4 Liquid egg to be used in food must be pasteurised except any liquid egg which is removed from the shell on the premises where the food is prepared and is either used forthwith or kept at a temperature not exceeding 50°F (10°C) and used within 24 hours.[1]

1 Liquid Egg (Pasteurisation) Regulations 1963/1503, as amended.

Meat

9.2.5 Raw and unprocessed meat may not contain any ascorbic acid, erythorbic acid, nicotinic acid, nicotinamide, nor any salt or other derivative of any of the foregoing substances.[1]

1 Meat (Treatment) Regulations 1964/19, as amended.

Milk-based drinks

9.2.6 Milk-based drinks must be heat treated by pasteurisation, sterilisation or the ultra high temperature method in accordance with prescribed methods.[1]

1 Milk-based Drinks (Hygiene and Heat Treatment) Regulations 1983/1508, as amended.

Cream

9.2.7 Cream, other than untreated cream, must be treated by pasteurisation, sterilisation or the ultra high temperature method in accordance with prescribed methods.[1]

1 Milk and Dairies (Heat Treatment of Cream) Regulations 1983/1509, as amended.

Animals and Fresh Meat

9.2.8 Animals to which there have been administered any hormonal substances other than permitted substances may not be sold or slaughtered for consumption.[1]

1 Animals and Fresh Meat (Hormonal Substances) Regulations 1988/849.

Milk

9.2.9 Appropriate heat treatments for milk sold as pasteurised, sterilised and ultra heat treated milk and appropriate coliform, plate count, phosphatase and turbidity tests are laid down, together with a system of licensing for milk producers and retailers and restrictions on the use of milk special designations.[1] Heat treatment requirements for semi-skimmed and skimmed milk are also prescribed.[2]

1 Milk (Special Designations) Regulations 1989/2383, as amended.
2 Milk and Dairies (Semi-skimmed and Skimmed Milk) (Heat Treatment and Labelling) Regulations 1988/2206, as amended.

Irradiation

9.2.10 From 1 January 1991 (13 February 1991 for poultry) certain fruits, vegetables, cereals, bulbs and tubers, spices and condiments, fish and shell fish, and poultry can be irradiated in Great Britain by prescribed methods; up to prescribed limits and for maximum periods of time subject to the irradiation premises being licensed. Where irradiation takes place in another member state, irradiated food can only be imported if the irradiation has taken place in a plant subject to official authorisation, recognition of which depends on the ministers being satisfied that the offical authorisation gives a standard of health protection equivalent to that given by an irradiation licence in Great Britain. Where irradiation takes place elsewhere overseas, food can only be imported if the exporting country is approved by the ministers and if they are satisfied that a licensing system exists in the exporting country which can ensure a measure of protection equivalent to an irradiation licence in Great Britain.

Records must be kept by licensees and special enforcement powers are provided for officers of the Minsters.[1]

1 Food (Control of Irradiation) Regulations 1990/2490.

9.3 Contaminants

9.3.1 *Any contaminant* which renders a foodstuff not of the nature or substance or quality demanded or breaches the food safety requirement[1] is an offence against the Food Safety Act 1990. Certain contaminants are controlled directly by regulations as follows.

1 Food Safety Act 1990, ss 8 or 14.

9.3.2 *Arsenic* levels in food are limited to 0.1 ppm in ready to drink beverages and 1.0 ppm in other foods. The limits do not apply in the case of fish or edible seaweed or any product containing such fish or seaweed for which a natural arsenic content of 1 ppm has been established. Higher limits are established for certain foods including onions, hops, liquorice, gelatine, yeast, chicory, herbs, spices, finings and clearing agents and specified chemicals.[1]

1 Arsenic in Food Regulations 1959/831, as amended.

9.3.3 *Lead* levels in food are limited to 1.0 mg/kg for all foods except where higher or lower limits are allowed.[1]

1 Lead in Food Regulations 1979/1254, as amended.

9.3.4 *Chloroform* may not be in any food sold in or imported into the United Kingdom save for any which may be controlled by the Medicines Act 1968 or any regulations made thereunder.[1]

1 Chloroform in Food Regulations 1980/36, as amended.

9.3.5 *Pesticides.* It is an offence to leave or cause to be left in any specified cereals, fruits and vegetables and certain products of animal origin (meat, milk, dairy produce and eggs) any residue of pesticide exceeding any maximum residue level laid down in the regulations[1] – see Chapter 15.

1 Pesticides (Maximum Residue Levels in Food) Regulations 1988/1378.

9.3.6 *Packaging Materials.* Food may be contaminated by the constituents of packaging materials with which its comes into contact. EC requirements (see Appendix B) are constantly changing and developing. Current law in Great Britain requires that materials and articles intended to come into contact with food must not, under foreseeable conditions of use, transfer their constituents to food in quantities which could endanger human health or bring about an unacceptable change in the nature, substance or quality of that food. For materials manufactured with vinyl chloride monomer or copolimers, limits are prescribed as to the quantity of vinyl chloride monomer which they may contain and to the quantity of vinyl chloride which they may transfer to food. For materials made from regenerated cellulose film, the substances from which they may be made are prescribed together with conditions and restrictions on their use.[1]

1 Materials and Articles in Contact with Food Regulations 1987/1523, as amended.

Chapter 10

The Food Safety Requirement and Dealing with Suspected Food

10.1 New Measures

10.1.1 The Food Safety Act 1990 introduced the concept of a 'food safety requirement'. It encompasses requirements as to food rendered injurious to human health and food which is unfit for human consumption, which existed in the Food Act 1984 and its predecessors, but introduced a new principle as to contaminated food. The food safety requirement was modelled on the general safety requirement for non-food products in Part II of the Consumer Protection Act 1987 but it is more precisely defined.

The Offence

10.1.2 It is unlawful to sell for human consumption, or offer, expose or advertise for sale for such consumption, or have in posession for the purpose of such sale or preparation for such sale, or to deposit with, or consign to, any other person for the purpose of such sale or preparation for such sale, any food which fails to comply with the food safety requirements.[1] It is important to distinguish the application of this offence from that of the sale to the prejudice of the purchaser of food which is not of the nature or substance or quality demanded[2] which was frequently used under former legislation to deal with cases of foreign bodies or mould in or on food. Under that offence it is necessary to prove a sale to the purchaser, thus restricting it to retail sales. This offence may arise at any point of manufacture or distribution.

1 Food Safety Act 1990, s 8 (1).
2 Food Safety Act 1990, s 14.

The Food Safety Requirement

10.1.3 Food fails to comply with the food safety requirement if –

(a) it has been rendered injurious to health by means of the addition of any article or substance to the food; the use of any article or substance as an ingredient in the preparation of the food; the abstraction of any constituent from food; or subjecting the food to any other process or treatment with the intention that it shall be sold for human consumption[1] (see **5.1.2**);
(b) it is unfit for human consumption;
(c) it is so contaminated (whether by extraneous matter or otherwise) that it would not be reasonable to expect it to be used for human consumption in that state.[2]

The term 'unfit for human consumption' has been well tried and tested under former food legislation. Extraneous objects, although undesirable, do not necessarily render food unfit for human consumption, but food which is putrid or decomposed is undoubtedy unfit. The courts have not always found it easy to decide, however, just how contaminated a food must be to render it unfit for human consumption. A pork pie bearing mould of the penicillin type, not of itself harmful to human beings, was unfit,[3] but a stale loaf of bread was unsuitable for human consumption but not necessarily unfit.[4] Potential toxicity from a foreign body would render food unfit, such as a dirty bandage in a load of bread,[5] but a piece of metal in a cream bun[6] or a piece of string in a loaf of bread[7] did not cause the food to be unfit. However, a prosecuting food authority need not be too concerned about these difficulties now because sub-paragraph (c) above should resolve them.

Because of the doubts about the extent of 'unfitness' many local authorities were accustomed to bringing proceedings for contaminated food as being 'not of the substance' demanded as discussed above. It was clear that there was a legislative gap between 'unfitness' on the one hand and 'not of the substance' on the other. During the debates on the Food Safety Bill it was proposed that the concept of 'unwholesomeness' would fill the gap. It has worked well in dealing with imported food[8] and there seemed to be no reason why it could not be similarly used in the food safety requirement.

The government believed, perhaps with good reason, that the term 'unwholesomeness' was itself capable of different interpretations and chose instead the wording in (c) above. It is an ingenious device which is already proving its worth. The full extent of the offence has yet to be tested in the courts but it does seem that the use of the words 'whether by extraneous matter or otherwise' when read with 'would not be reasonable to expect it to be used for human consumption in that state' would bring within the offence most occurrences which might deter the ordinary consumer from eating a food.

Due regard must be had to the words 'in that state' because there would appear to be no offence if the food was to be subjected to further processing or treatment.[9]

1 Food Safety Act 1990, s 7 (1).
2 Food Safety Act 1990, s 8 (2).
3 *David Greig Ltd v Goldfinch* (1961) 105 Sol Jo 367, DC.
4 *Kyle v Laird* 1951 JC 65.
5 *Chibnall's Bakeries v Cope Brown* [1956] Crim LR 263; *Meah v Roberts; Lansley v Roberts* [1978] 1 ALL ER 97, [1977] 1 WLR 1187.
6 *J Miller Ltd v Battersea Borough Council* [1956] 1 QB 43, [1955] 3 All ER 279.
7 *Turner & Son Ltd v Owen* [1956] 1 QB 48, [1955] 3 All ER 565.
8 Imported Food Regulations 1984/1918, as amended, reg 6 (1) (c).
9 *R v Archer, ex p Barrow, Lane & Ballard Ltd* (1983) 147 JP 503, CA.

Treatment of Batches, Lots or Consignments of Food

10.1.4 Where any food which fails to comply with the food safety requirements is part of a batch or lot or consignment of food of the same class or description it is presumed, until the contrary is proved, that all the food in that batch. lot or consignment fails to comply with those requirements.[1] The burden of proof that the whole of a batch, lot or consignment of food does not fail to satisfy the food safety requirement is on the owner of the food and not the investigating officer. In circumstances where a part of such batch, lot or consignment is obviously in breach of the requirements, that would be a difficult burden to discharge.

1 Food Safety Act 1990, s 8 (3).

Unfit Meat

10.1.5 The Act creates a further presumption that any product or part of a product derived wholly or partly from an animal which has been slaughtered in a knacker's yard, or of which the carcase has been brought into a knacker's yard (in Scotland, otherwise than in a slaughterhouse), is unfit for human consumption.[1]

1 Food Safety Act 1990, s 8 (4).

Consumer Complaints

10.1.6 Offences against the food safety requirement are most likely to arise from consumer complaints to local environmental health departments. Guidance to food authorities on the handling of consumer complaints has been given by the ministers[1] and includes the following important points:

(a) *Notification of complaints*: As a general rule any person who may be prosecuted as a result of a consumer complaint should be notified that the complaint has been made as soon as reasonably practicable. Normally, as soon as preliminary investigations suggest that the complaint may be well founded, the food authority should notify the supplier, manufacturer or importer. The initial notification may be oral and should be followed by notification in writing. The written notification should describe the date and nature of the complaint. Other potential defendants should be notified as they emerge. Exceptionally there may be circumstances in which notification could impede the progress of further investigations. In these circumstances, notification should take place as soon as can be achieved without prejudicing further investigations.

(b) *Involvement of other food authorities*: If an investigation of a complaint brings to light a problem or potential problem outside the area of the enforcing authority, the other appropriate authorities directly affected (for example the authority where the food was manufactured and/or the authority where a company's decision-making base is located) should be informed as soon as possible. If there is a public health problem which may affect a wide geographical area, the Department of Health, the Ministry of Agriculture, Fisheries and Food, the Scottish Office or the Welsh Office should be informed, as a hazard warning or emergency order may be needed.

(c) *Investigation of complaint samples*: The authorised officer will need to consider whether any scientific investigation is to be undertaken. If the authorised officer is in any doubt he should consult the public analyst and/or food examiner who will advise on the form of scientific investigation which may be appropriate for that complaint sample, particularly since some will require a combination of analysis and examination. If the authorised officer considers that a complaint sample should be analysed, he should send it to a public analyst. If he considers it should be microbiologically examined, he should send it to a food examiner. If he considers any other investigation is necessary, he should ensure that the person to whom he sends the complaint sample is suitably qualified and able to give evidence in the event of a prosecution.

(d) *When sample can be surrendered*: The food authority should try to comply with any reasonable request provided that it does not impede the proper storage, analysis, examination or evidential value of the samples.

(e) *Taking legal proceedings*: Prosecutions should be brought without unnecesary delay. In Scotland, the decision on referral to the Prosecutor Fiscal should be taken without unnecessary delay.

1 Code of Practice No 2: Legal Matters (HMSO) (ISBN 0 113 21353 0).

10.2 Inspection and Seizure of Suspected Food

10.2.1 The Food Safety Act 1990 has re-enacted the provisions of the Food Act 1984 and its predecessors as to the power of authorised officers to seize and detain suspect food and, on the order of a justice of the peace, have it destroyed or disposed of to prevent it being used for human consumption.[1] However, the powers have been considerably extended and improved and are now based on the food safety requirements discussed above in addition to their application to foods likely to cause food poisoning or any communicable disease.

1 Food Safety Act 1990, s 9 (1), (2) and (3).

The Powers

10.2.2 An authorised officer of a food authority may at all reasonable times inspect any food intended for human consumption which has been sold or is offered or exposed for sale or is in the possession of, or has been deposited with or consigned to, any person for the purpose of sale or preparation for sale. Where it appears to the officer that the food fails to comply with the food safety requirement or it is likely to cause food poisoning or any disease communicable to human beings, he may either give notice to the person in charge of the food that, until the notice is withdrawn, the food or any specified portion of it is not to be used for human consumption and either is not to be removed or is not to be removed except to some place specified in the notice, or he may seize the food and remove it in order to have it dealt with by a justice of the peace.

The form of notice is prescribed[1] and it is an offence to knowingly contravene the requirements of a notice. The inclusion of the word 'knowingly' in this offence requires proof by the prosecution of guilty knowledge or intent; that is, that the defendant knew of the existence of the notice and its contents.[2] However, in this case the notice would have to be served on the person in charge of the food and it is therefore submitted that proof of service of the notice in accordance with the Act[3] would be sufficient since the recipient, being aware of the inspection of the food, could hardly claim to be unaware of its contents.

Under former legislation the term 'all reasonable times' was assumed to mean 'during normal business hours' – see a discussion of inspection powers in Chapter 17. However, the meaning should be considered in the light of the mischief intended to be dealt with and it is submitted that the protection of the public against dangerous food is so overwhelmingly important that authorised officers should not be deterred from exercising these powers at any time of the night or day until there is judicial precedent to the contrary.

For similar powers in relation to imported food see Chapter 14.

1 Detention of Food (Prescribed Forms) Regulations 1990/2614.
2 *Woolmington v DPP* [1935] AC 462, 25 Cr App Rep 72, HL; and *Gaumont British Distributors Ltd v Henry* [1939] 2 KB 711, [1939] 2 All ER 808.

3 Food Safety Act 1990, s 50.

Termination of Notice

10.2.3 Where an authorised officer has issued a notice he must, as soon as is reasonably practicable and in any event within 21 days, determine whether or not he is satisfied that the food complies with the food safety requirements. If he is so satisfied he must forthwith withdraw the notice. If he is not so satisfied, he must seize the food and remove it to have it dealt with by a justice of the peace.[1] A notice withdrawing a detention notice should, notwithstanding the above, be served as soon as possible to avoid further deterioration of the food.[2]

1 Food Safety Act 1990, s 9 (4).
2 Code of Practice No 4: Inspection, Detention and Seizure of Suspect Food, para 28 (HMSO) (ISBN 0 11 321350 6).

Referral to a Justice of the Peace

10.2.4 Where the authorised officer exercises his powers to refer the matter to a justice of the peace he must inform the person in charge of the food of his intention so to do. A person who might be liable for prosecution in respect of the food concerned must be given an opportunity to attend before the justice of the peace, to give evidence and to call witnesses.[1]

If it appears to the justice of the peace, on the basis of such evidence as he considers appropriate in the circumstances, that the food fails to comply with the food safety requirement he must condemn the food and order the food to be destroyed or to be so disposed of as to prevent it from being used for human consumption; and any expenses reasonably incurred in connection with the destruction or disposal must be defrayed by the owner of the food.[2]

Referral to a justice of the peace should normally be within two days of seizure and in the case of perishable food such referral should be as soon as possible.[3]

1 Food Safety Act 1990, s 9 (5).
2 Food Safety Act 1990, s 9 (6).
3 Code of Practice No 4: Inspection, Detention and Seizure of Suspect Food, para 11.

Compensation

10.2.5 If a notice issued by an authorised officer is withdrawn or if a justice of the peace refuses to condemn the food the food authority must compensate the owner of the food for any depreciation in its value resulting from the action taken by the authorised officer. Any dispute as to the amount of compensation must be settled by arbitration.[1]

1 Arbitration Acts 1950, 1975 and 1979 – see 2 Halsbury's Statutes (4th Edn) 535, 603 and 608.

Prosecution

10.2.6 Action by an authorised officer under the above provisions does not preclude a subsequent prosecution by the food authority in respect of the failure of the food safety requirement or the food having been rendered injurious to health. An order by a justice of the peace for the food to be destroyed or disposed of is sufficient evidence of the failure of the food to comply with the food safety requirements.[1]

1 Food Safety Act 1990, s 9 (9) (c).

Action in Cases of Food Poisoning

10.2.7 In the event of a food poisoning outbreak, the authorised officer should seek expert advice from the consultant in communicable disease control (CCDC) and may also need to contact the local public health laboratory service (PHLS) or the public analyst. In Scotland advice should be sought, for example, from the consultant in public health medicine or the microbiologist in charge of the laboratory investigating the outbreak. Whenever possible and appropriate there should be full and open discussions with the owner or person in charge of the food and with the manufacturer.[1]

1 Code of Practice No 4: Inspection, Detention and Seizure of Suspect Food, para 14.

Security of Detained Food

10.2.8 If the authorised officer has any doubts about the security or physical care of the food, the detention of food notice should specify a place to which the food should be removed. If the food is removed to another authority's area the officer should notify that authority and make any necessary arrangements for the food to be checked while it is being detained. In all cases, especially with highly perishable food, the officer should act expeditiously at every stage and provide full information to those required to carry out analysis or examination of samples of the food.

If food is detained where it is found, the authorised officer should be satisfied that adequate arrangements can be made to ensure its security and prevent tampering. This could be by locking the food in a room or enclosure, sealing it or marking it in such a way that removal of part is obvious and/or by requiring the continuous presence of security personnel. The officer should organise periodic monitoring of the food throughout the period of detention. Before making such arrangements regard should be had to the nature of the food, the quantity, any health hazard that it represents and the ownership of the premises where it is located. The officer should generally avoid leaving it in the charge of, or in premises owned by, any person who may be prosecuted for an offence concerning food injurious to health or in breach of the food safety requirements.[1]

1 Code of Practice No 4: Inspection, Detention and Seizure of Suspect Food, paras 14, 16, 17.

Destruction or Disposal of Food

10.2.9 Where destruction or disposal of food is necessary the food authority is responsible for the necessary arrangements. The food should be fully supervised until it can be dealt with in the appropriate manner. If possible, and if there is likely to be considerable delay before destruction, the food should be disfigured in some way to prevent any possibility of it being returned to the market place illegally or accidentally.

If food is to be destroyed, the food authority should arrange for total destruction of the food, for example by incineration. If this is not possible the authority should arrange for such a degree of disfigurement that the food could never again enter the food chain, for example by flattening tin cans for disposal in a suitably licensed land-fill site.[1]

1 Code of Practice No 4: Inspection, Detention and Seizure of Suspect Food, paras 52, 53.

Wider Implications

10.2.10 The provisions discussed above are intended and suitable for dealing with individual batches, lots or consignments of dangerous food disovered and dealt with within the area of a food authority. For cases with wider implications the powers of food authorities in relation to emergency prohibition notices[1] and those of the ministers in relation to emergency control orders[2] should be borne in mind (see Chapter 11).

1 Food Safety Act 1990, s 12.
2 Food Safety Act 1990, s 13.

Chapter 11

Executive Notices, Court Orders and Prohibitions

11.1 Introduction

Executive Powers

11.1.1 The Food Safety Act 1990 introduced some extensive new powers to enable food authorities and the ministers to deal effectively with unsafe food. Although based to some extent on precedents created by the Health and Safety at Work Act 1974 and the Food and Environment Protection Act 1985 the powers, especially those conferred on environmental health officers, place an awesome responsibility on the officers concerned. Failure to use the powers effectively in circumstances resulting in serious damage to public health, or misguided or excessive use of them incurring unnecessary costs to the food industry, could have serious consequences for the employing food authority or the ministers.

Codes of Practice

11.1.2 Conscious of the need to coordinate, guide and direct the work of enforcement officers, the Ministers have exercised their powers[1] to produce statutory codes of practice with respect to improvement notices, prohibition procedures and hygiene inspections. These codes are very detailed and although they are not directly binding on food authorities, the ministers are empowered to issue directions requiring an authority to take any specified steps to comply with a code. Such directions are enforceable by *mandamus*[2] in England and Wales or by an order of the Court of Session in Scotland.[3] Ministers are obliged to consult with interested organisations before making a code and it is, by implication, open to persons in the food industry to complain to the ministers if guidance in a code is not observed.

1 Food Safety Act 1990, s 40.
2 *Mandamus* – a remedial order issued to a person or body corporate by the Queen's Bench Division of the High Court, generally used to compel public officials to fulfil a statutory duty.
3 An order under s 45 of the Court of Session Act 1988.

Default Powers

11.1.3 Where the minister is satisfied that a food authority has failed to discharge any duty imposed by the Food Safety Act 1990 or any regulations made or continued in force thereunder, and the failure affects the general interests of consumers, he may by order empower another food authority, or one of his officers, to discharge that duty in place of the authority in default. In determining whether this power should be exercised the minister may, if he thinks fit, cause a local inquiry to be held under the relevant provisions of the Local Government Act 1972.[1]

Ministers may not exercise their default powers where the alleged default of a food authority adversely affects the interests only of a food business and not consumers generally. Further, officers of food authorities are not personally liable in respect of anything done by them in the execution of the Act and within the scope of their employment and they may be indemnified, subject to certain conditions, by their employing food authority for any damages which may be awarded against them.[2] However, the food authority itself is liable in respect of the actions of its officers and its actions may be the subject of judicial review.

1 Local Government Act 1972, s 250 (2)–(5); Local Government (Scotland) Act 1973, s 210 (3)–(8).
2 Food Safety Act 1990, s 44 (1) and (3).

11.2 Improvement Notices

The Power to Issue an Improvement Notice[1]

11.2.1 An authorised officer of an enforcement authority may, if he has reasonable grounds for believing that the proprietor of a food business is failing to comply with any hygiene regulations, or regulations controlling the processing or treatment of food, issue a notice (called an 'improvement notice'). The notice must state the officer's grounds for believing that the proprietor is failing to comply with the regulations; specify the matters which constitute the proprietor's failure so to comply; specify the matters which, in the officer's opinion, the proprietor must take in order to secure compliance; and require the proprietor to take those measures, or measures which are at least equivalent to them, within such period (not less than 14 days) as may be specified in the notice.

This provision is modelled on s 21 of the Health and Safety at Work Act 1974[2] and, like that section, requires the officer to decide whether or not regulations are being breached. There is authority for saying that a public official, acting in good faith, is the sole judge of necessity or expediency in the interests of public health or consumer protection and cannot be challenged as to an opinion honestly held.[3] Nevertheless, the officer is required to exercise a quasi-judicial function before issuing an improvement notice and must, therefore, be thoroughly familiar with the relevant regulations and capable of interpreting them fairly and correctly.

1 Food Safety Act 1990, s 10.
2 Health and Safety at Work Act 1974; 19 Halsbury's Statutes (4th Edn) 620.
3 *Robinson v Sunderland Corpn* [1899] 1 QB 751; *R v Comptroller General of Patents, ex p Bayer Products Ltd* [1941] 2 KB 306, [1941] 2 All ER 677, CA; *Point of Ayr Collieries Ltd v Lloyd-George* [1943] 2 All ER 546, CA; *Re City of Plymouth (City Centre) Declaratory Order 1946, Robinson v Minister of Town & Country Planning* [1947] KB 702, [1947] 1 All ER 851, CA; *Smith v East Alloe RDC* [1956] AC 736, [1956] 1 All ER 855, HL; *Ross-Clunis v Papadopoullos* [1958] 2 All ER 23, [1958] 1 WLR 546, PC; *Customs and Excise Comrs v Cure and Deeley Ltd* [1962] 1 QB 340, [1961] 3 All ER 641.

The Relevant Regulations

11.2.2 There is no authoritative list. The test is whether the regulations make provision for requiring, prohibiting or regulating the use of any process or treatment in the preparation of food; or for securing the observance of hygienic conditions and practices in connection with the carrying out of commercial operations with respect to food or food sources.[1] The majority of circumstances requiring the issue of an improvement notice fall under the food hygiene regulations (printed in bold type below). A non-exhaustive list of regulations is as follows:[2]

Public Health (Shell-fish) Regulations 1934
Milk and Dairies (General) Regulations 1959
Ice-Cream (Heat Treatment etc) Regulations 1959
Food Hygiene (Docks, Carriers etc) Regulations 1960
Liquid Egg (Pasteurisation) Regulations 1963
Meat (Treatment) Regulations 1964
Food Hygiene (Markets, Stalls and Delivery Vehicles) Regulations 1966, as amended in 1990 and 1991
Food Hygiene (General) Regulations 1970, as amended in 1990 and 1991
Poultry Meat (Hygiene) regulations 1976
Drinking Milk Regulations 1976
Slaughterhouses (Hygiene) Regulations 1977
Meat (Sterilisation and Staining) Regulations 1982
Milk-Based Drinks (Hygiene and Heat Treatment) Regulations 1983
Milk and Dairies (Heat Treatment of Cream) Regulations 1983
Fresh Meat Export (Hygiene and Inspection) Regulations 1987
Milk and Dairies (Semi-skimmed and Skimmed Milk) (Heat Treatment and Labelling) Regulations 1988
Bovine Offal (Prohibition) Regulations 1989
Milk (Special Designations) Regulations 1989
Ungraded Eggs (Hygiene) Regulations 1990

Note: This list will be considerably extended when regulations to implement impending EC measures on health and hygiene problems in meat, meat products, milk and milk products and egg products are made. For SI numbers and relevant amendments to the above regulations see Appendix A. For a discussion of food hygiene law see Chapter 12.

1 Food Safety Act 1990, s 10 (3).
2 Code of Practice No 6: Prohibition Procedures – the schedule.

The purpose of Improvement Notices

11.2.3 Improvement notices are intended to deal with situations where there is a breach of the relevant regulations, but such breach is not sufficiently serious to justify emergency action (see **11.4** below) but is too immediate in its impact to await a prosecution. Issue of an improvement notice does not, of course, preclude a prosecution in relation to the same incident. Breach of an improvement notice is itself an offence. The following are examples of circumstances where use of an improvement notice would be appropriate:[1]

- where rodent proofing or the provision of a flying insect screen is necessary to prevent any risk of infestation;
- where there are deficiencies in the structure or facilities in the food business, eg where structural repairs are required or where additional equipment is necessary to comply with temperature control requirements;
- where facilities for washing food, equipment and utensiles are inadequate;

- where there is inadequate ventilation in a kitchen area;
- where there is inadequate artificial lighting in a food room;
- where there is a failure to maintain the premises in a satisfactory state of cleanliness.

1 Code of Practice No 5: The Use of Improvement Notices, para 8.

11.2.4 *The notice* should be in the prescribed form[1] and should contain all necessary information, with references as to the regulations which are allegedly contravened; as to the matters which do not comply; and how such compliance can be achieved. In the event of non-compliance with an improvement notice a prosecution may follow and a notice which was wrongly drafted would be fatal to the proceedings. Service of the notice should comply with statutory requirements.[2] The minimum time which may be specified in an improvement notice is 14 days. The period of time given for remedial work to be carried out should be realistic in the light of the amount of work and the mischief to be corrected. Although there is no provision in the Act for extensions of time, in certain circumstances it may be reasonable to defer enforcement of an improvement notice until a further period of time has elapsed.

One problem in drafting an improvement notice is that the food hygiene regulations are capable of widely different interpretations leaving much to the discretion of enforcement officers. Officers should take care to ensure that their reasons for issuing an improvement notice are based on sound hygiene practice.

An improvement notice must be served on the proprietor of the food business. If that proprietor is a body corporate, the notice should be addressed to the company secretary at the company's registered office. It would not be sufficient to address it to or leave it with, for example, a store manager.

1 Food Safety (Improvement and Prohibition – Prescribed Forms) Regulations 1991/100.
2 Food Safety Act 1990, s 50.

Appeals

11.2.5 A person who is aggrieved by a decision of an authorised officer to serve an improvement notice may appeal to a magistrates' court or, in Scotland, to the sheriff.[1] The effect of an appeal is to stop time running on the notice. If an improvement notice is, in the opinion of the food business proprietor, issued wrongly or unreasonably he should first try to negotiate a solution to the problem with the issuing officer or his superiors. If that is unsuccessful an appeal should be lodged with the clerk to the justices without delay.

1 Food Safety Act 1990, s 37 (1) (a).

11.3 Prohibition Orders

11.3.1 New powers were conferred on the courts by the Food Safety Act 1990 to impose certain prohibitions on food business proprietors and managers after conviction for certain offences. These powers replace those to close food premises in former legislation but go much further. They include, for example, a prohibition on a proprietor or manager from participating in the management of any food business for an indefinite period. Bearing in mind that these powers may be exercised by a magistrates' court, subject of course to appeal, they are dramatic indeed. They were thought necessary to prevent such persons from perpetuating dangerous trading practices and from closing down a business after conviction and prohibition only to open another business in a different area.

Prohibition Powers[1]

11.3.2 The courts are empowered to make prohibition orders of two classes. They are:

(a) where a food business proprietor is convicted of an offence under any of the relevant regulations (see **11.2.2** above) and the court by or before which he is so convicted is satisfied that the health risk condition is fulfilled with respect to that business. The health risk condition is fulfilled if any of the following involves risk of injury to health –
   (i) the use for the purposes of the business of any process or treatment – the court may prohibit the use of the process or treatment for the purposes of the business;
   (ii) the construction of any premises used for the purposes of the business, or the use for those purposes of any equipment – the court may prohibit the use of the premises or equipment for the purposes of the business or any other food business of the same class or description; and
   (iii) the state or condition of any premises or equipment used for the purposes of the business – the court may prohibit the use of the premises or equipment for the purposes of any food business.

(b) where a food business proprietor or a manager of a food business is convicted of an offence under any of the regulations in paragraph **11.2.2** which are for securing the observance of hygienic conditions and practices, and the court by which or before which he is so convicted thinks it proper to do so in all the circumstances of the case, the court may by an order prohibit the proprietor or manager from participating in the management of any food business, or any food business of a class or description specified in the order.

1 Food Safety Act 1990, s 11.

Managers

11.3.3 For the purposes of the power under (b) above, a manager in relation to a food business means any person who is entrusted by the proprietor with the day to day running of the business, or any part of the business. Before a manager can be the subject of a prohibition order he must himself have been convicted of a relevant offence. There is a popular and erroneous belief that a manager by whose act or default an offence by the proprietor is alleged to have been committed[1] can, on the conviction of the proprietor, be the subject of a prohibition order. It is, of course, open to a prosecuting food authority, knowing that the proprietor will plead a defence of act or default of his manager, to bring proceedings against that manager in addition to or instead of the proprietor, thus enabling the court, if it thinks fit, to issue a prohibition order against the manager.

1 Food Safety Act 1990, s 20.

Application for a Prohibition Order

11.3.4 The Act does not stipulate the steps which should precede an inviation to a court to make a prohibition order. Important advice is given to food authorities[1] which suggests that the premises subject to a prosecution should be inspected prior to the hearing, and if the enforcement officer believes that the defect giving rise to the prosecution either has not been removed or has been removed but has re-occurred because of a lapse in management, the court should be invited to make the necessary order after the conviction has been secured. Consistent with natural justice and current legal requirements[2] the officer should inform the proprietor of his intention to invite the court to make a prohibition order and any evidence available to the food authority should be made available to the proprietor or his solicitors prior to the hearing.

1 Code of Practice No 6: Prohibition Procedures, para 53.
2 Magistrates' Courts (Advance Information) Rules 1985/601.

Service of Prohibition Orders

11.3.5 As soon as practicable after the making of a prohibition order the enforcement authority must serve[1] a copy of the order on the proprietor of the business (and/or the manager as appropriate), and in the case of a class (a) order (see above) affix a copy of the order in a conspicuous position on such premises used for the purposes of the business as they consider appropriate.

The food authority should consider the desirability of asking the court to include a provision in it that it should not be defaced or removed.[2] If that is done and the order is then defaced or removed an offence would have been committed. It has been suggested[3] that removal or defacement of an order would be criminal damage because the order is the property of the food authority. It is not, it is suggested, a course likely to achieve the desired result, because there is an element of guilty knowledge or recklessness in the offence which may be difficult to prove.

1 Food Safety Act 1990, s 50.
2 Magistrates' Courts Act 1980, s 63.
3 Code of Practice No 6: Prohibition Procedures; para 72; Criminal Damage Act 1971, s 1.

Cessation of a Prohibition Order

11.3.6 A prohibition order ceases to have effect in a case under **11.3.2** (a) above on the issue by an enforcement authority of a certificate to the effect that they are satisfied that the proprietor has taken sufficient measures to secure that the health risk condition is no longer fulfilled with respect to the business, and in a case under **11.3.2** (b) above on the giving by the court of a direction to that effect. The certificate must be issued within three days of the authority being satisfied as mentioned above.

If the proprietor applies to the enforcement authority for such a certificate the authority must determine, as soon as it is reasonably practical and in any event within 14 days, whether or not they are so satisfied. If they determine that they are not so satisfied they must give notice to the proprietor of the reasons for that determination.

In the case of an order under **11.3.2** (b) above the proprietor or manager must apply for cessation to the court himself. If the court thinks it proper, having regard to all the circumstances and the conduct of the proprietor or manager since the making of the order, it must give a direction for cessation. No such application may be entertained within six months after the making of the prohibition order or within three months after the making by the proprietor or manager of a previous application for such a direction.

Appeals

11.3.7 Any person who is aggrieved by a decision of a magistrates' court may appeal to the Crown Court.[1]

1 Food Safety Act 1990, s 38(b).

11.4 Emergency Prohibition Notices and Orders

11.4.1 The Food Safety Act 1990[1] provides the means for an enforcement officer to initiate action leading to a prohibition order, by the service of an emergency prohibition notice. Although such a notice must be confirmed by a court by the issue of an emergency prohibition order within three days of the issue of the notice, the effects of such a notice on a food business are serious and the burden of responsibility placed on the officer is a heavy one. Code of Practice No 6: Prohibition Procedures,[2] contains detailed recommendations for the guidance of food authorities and their authorised enforcement officers which should be followed closely.

1 Food Safety Act 1990, s 12.
2 Code of Practice No 6: Prohibition Procedures (HMSO ISBN 0 11 321349 2).

The Health Risk Condition

11.4.2 If an authorised officer is satisfied[1] that the health risk condition is fulfilled with respect to a food business he may, by a notice[2] served on the proprietor of the business, impose the appropriate prohibition. The health risk condition is that discussed in **11.3.2** above but with the difference that the risk of injury to health is *imminent*. There appear to be no relevant authorities on the legal meaning of the term 'imminent'. It is suggested that its ordinary meaning of 'impending threateningly; hanging over one's head; ready to overtake one; coming on shortly',[3] is apposite and its meaning should not be restricted to 'immediate'. For example, if the condition of premises appears to carry a high risk of causing an outbreak of food poisoning within the next few days, an authorised officer would be entitled to issue an emergency prohibition notice. If the risk was honestly believed to be imminent in any circumstances to a qualified and experienced environmental health officer that belief ought not to be questioned by a court. However, see the provisions as to compensation below.

1 *Robinson v Sunderland Corpn* [1899] 1 QB 751; *R v Comptroller General of Patents, ex p Bayer Products Ltd* [1941] 2 KB 306, [1941] 2 All ER 677, CA; *Point of Ayr Collieries Ltd v Lloyd-George* [1943] 2 All ER 546, CA; *Re City of Plymouth (City Centre) Declaratory Order 1946, Robinson v Minister of Town & Country Planning* [1947] KB 702, [1947] 1 All ER 851, CA; *Smith v East Alloe RDC* [1956] AC 736, [1956] 1 All ER 855, HL; *Ross-Clunis v Papadopoullos* [1958] 2 All ER 23, [1958] 1 WLR 546, PC; *Customs and Excise Comrs v Cure and Deeley Ltd* [1962] 1 QB 340, [1961] 3 All ER 641.
2 Food Safety (Improvement and Prohibition – Prescribed Forms) Regulations 1991/100.
3 *Oxford English Dictionary.*

Display of Notice

11.4.3 A soon as practicable after the service of an emergency prohibition notice, the enforcement authority must affix a copy of the notice in a conspicuous position on such premises used for the purpose of the business as they consider appropriate – see **11.3.6** above.

Contravention of an Emergency Prohibition Notice or Order

11.4.4 It is an offence for any person to *knowingly* contravene an emergency prohibition notice or order. This offence requires proof of guilty intent and the authorised officer should establish that the alleged offender knew of the existence of the notice and understood its contents. See Chapter 17 for the investigation of offences.

Applying for an Emergency Prohibition Order

11.4.5 Unless an application is made to a court within three days beginning with the service of the emergency prohibition notice, such notice ceases to have effect. At least one day before the application is made to the court a notice[1] of that intention must be served[2] on the proprietor of the business.

1 Food Safety (Improvement and Prohibition – Prescribed Forms) Regulations 1991/100.
2 Food Safety Act 1990, s 50.

Service of an Emergency Prohibition Order

11.4.6 As soon as practicable after the making of an emergency prohibition order, the enforcement authority must serve[1] a copy of the order on the proprietor of the business and affix a copy of the order in a conspicuous position on such premises used for the purposes of that business as they consider appropriate. See **11.3.5** above.

1 Food Safety Act 1990, s 50.

Cessation of Emergency Prohibition Notices and Orders

11.4.7 An emergency prohibition notice ceases to have effect if no application is made to a court for an emergency prohibition order within a period of three days beginning with the service of the notice; or, if such an application is so made, on the determination or abandonment of the application.

An emergency prohibition notice or order ceases to have effect on the issue by the enforcement authority of a certificate[1] to the effect that they are satisfied that the proprietor has taken sufficient measures to secure that the health risk condition is no longer fulfilled with respect to the business. The certificate must be issued within three days of the authority being so satisfied. If an application for a certificate is made by the proprietor of the business the authority must determine as soon as is reasonably practicable, and in any event within 14 days, whether or not they are so satisfied and ,if they determine that they are not so satisfied, give notice[1] to the proprietor of the reasons for that determination.

1 Food Safety (Improvement and Prohibition – Presscribed Forms) Regulations 1991/100.

Compensation

11.4.8 Where an emergency prohibition notice has been served on the proprietor of a business, the enforcement authority must compensate him in respect of any loss suffered by reason of his complying with the notice unless an application for an emergecy prohibition order is made within three days beginning with the service of the notice and the court declares itself satisfied, on the hearing of the application, that the health risk condition was fulfilled with respect to the business at the time when the notice was served.

Any dispute as to compensation must be determined by arbitration[1] or, in Scotland, by a single arbitrater appointed, failing agreement between the parties, by the sheriff.

1 Arbitration Acts 1950, 1975 and 1979 – see 2 Halsbury's Statutes (4th Edn) 535, 603 and 608.

11.5 Emergency Control Orders

11.5.1 The Food Safety Act 1990 empowers the minister to make emergency control orders where there is an imminent risk of injury to health. These powers are not confined to individual food businesses but may apply generally as appropriate. Their purpose is to deal with broadly-based threats to health beyond the scope of locally-based enforcement authorities.

The Power to Make an Emergency Control Order[1]

11.5.2 If it appears to the Minister of Agriculture, Fisheries and Food, the Secretary of State or the Secretary of State for Wales ('the Minister'), that the carrying out of commercial operations with respect to food, food sources or contact materials of any class or description involves, or may involve, imminent risk of injury to health, he may, by order (known as an emegency control order) prohibit the carrying out of such operations with respect to food, food sources or contact materials of that class or description. It is an offence to knowingly contravene an emergency control order.

1 Food Safety Act 1990, s 13.

Exceptions and Directions

11.5.3 The minister has the power to consent, conditionally or unconditionally, to the doing of something in a particular case which is prohibited by an emergency control order. This is to provide a necessary degree of flexibility to allow, for example, the movement of prohibited food from one place to another to achieve the objectives of an emergency control order. Where such consent is given it is not an offence for any person to comply therewith notwithstanding the fact that the action may be contrary to the emergency control order.

Power also exists for directions to be given for the purpose of preventing the carrying out of commercial operations in accordance with an emergency control order. It is an offence to fail to comply with a direction so given. The minister may himself may do anything which appears to him to be necessary or expedient for the purposes of an emergency control order. In the event that the minister does anything under this power as a consequence of any person failing to comply with an emergency control order he may recover from that person any expenses reasonably incurred by him in that connection.

11.6 Other Emergency Orders

Emergency Orders[1]

11.6.1 Where there exist circumstances which are likely to create a hazard to human health through human consumption of food, the Minister of Agriculture, Fisheries and Food or the Secretary of State may make an order by statutory instrument designating any area of land in the United Kingdomn, or sea within the British fishery limits, or both such area of land and sea from which anything is or may be derived which is or may become unfit for human consumption. The order may contain emergency prohibitions and it is an offence to contravene an emergency order or to cause or permit any other person to do so.

1 Food and Environment Protection Act 1985, as amended, s 1.

Scope of Emergency Orders

11.6.2 In its original form this power was confined to circumstances where there had been an escape of contaminants to food. In its amended form[1] there is no reference to 'escape' thus providing wider application of the power and greater flexibility in its operation. For example, the power was widely used during 1991 to prohibit the taking of bivalve molluscs and certain crabs from certain waters around northern Scotland because of suspected paralytic shellfish poisoning. It would have been difficult to use the power in its original form for this purpose because there was no positive proof that there had been an 'escape'. On the other hand, the power in its original form was widely used following the Chernobyl nuclear accident in the Soviet Union when sheep in Scotland, northern England and Wales were contaminated by radioactivity.

1 Food Safety Act 1990, s 51 (1) and (2).

Chapter 12

Food Hygiene

12.1 Introduction

12.1.1 Basic food hygiene in the selling, keeping, storage and transport of food for human consumption is subject to three principal sets of regulations. They are the Food Hygiene (Docks, Carriers etc) Regulations 1960 (SI 1960/1602); the Food Hygiene (Markets, Stalls and Delivery Vehicles) Regulations 1966 (SI 1966/791) and the Food Hygiene (General) Regulations 1970 (SI 1970/1172). All have been amended a number of times (see Appendix A).

These regulations are overdue for review and updating but have generally worked well and have established standards of hygiene in the United Kingdom which are comparable with the best in any other member state of the EC. The regulations are logically constructed to ensure:

(a) the cleanliness of premises, vehicles and equipment used for the purposes of a food business;
(b) the hygienic handling of food;
(c) the cleanliness of persons engaged in the handling of food and of their clothing, and the action to be taken when they suffer from or are the carriers of certain infections likely to cause food poisoning;
(d) the construction of premises, vehicles and ships used for the purposes of a food business; their repair and maintenance;
(e) the provision of water supply and washing facilities;
(f) the proper disposal of waste material;
(g) the temperatures at which certain foods are to be kept.

Because there can be no absolute standards of cleanliness the regulations are necessarily broadly drafted leaving a great deal to the common sense of enforcement officers and the courts. In the past this has resulted in widely differing interpretations between food authorities which still exist to a lesser extent today. However, enforcement is now much improved as a result of better coordination of enforcement and statutory codes of practice issued under the Food Safety Act 1990.[1] The regulations have also acquired a new importance because the issue of improvement notices[2] and emergency prohibition notices[3] is triggered, in part, by an alleged breach of the regulations (see Chapter 11).

1 Code of Practice No 9: Food Hygiene Inspections (HMSO ISBN 0 11 321467 7); Code of Practice

No 10: Enforcement of the Temperature Control Requirements of Food Hygiene Regulations (HMSO ISBN 0 11 321465 0).
2 Food Safety Act 1990, s 10.
3 Food Safety Act 1990, s 12.

Future Developments

12.1.2 In 1989 a proposal for a Council regulation on the protection of human health by monitoring the wholesomeness of foodstuffs and observing basic principles of hygiene during their production, handling and sale, was issued for consultation. The new directive should have been adopted by the end of 1992 and may be implemented by regulations in the United Kingdom by the end of 1993. The approach will differ from that of the above regulations in that there will be greater emphasis on the monitoring of hygiene standards by food traders in coordination with food authorities, backed by more stringent standards of hygiene with respect to premises, equipment and staff.

Registration or Licensing of Food Premises

12.1.3 The ministers are empowered to make regulations[1] requiring the licensing or registration of food premises. On 17 December 1991 the Department of Health announced that regulations[2] covering England, Scotland and Wales will come into force on 1 February 1992. From that date any person operating a food business will have three months in which to register their premises. From 1 May 1992 it will be an offence to operate a food business from unregistered premises. The registration authority will be district councils and the work will be done by environmental health officers. They have no discretion to refuse registration.

The purpose of registration is to provide information to the enforcement authorities. It is not a licensing system and registration does not itself affect the fact that all food premises may be inspected.

The regulations include exemptions from the requirement to register, including exemptions for any premises which are used on less than five days in five consecutive weeks, domestic premises where bed and breakfast accomodation is provided in not more than three bedrooms, and premises used by voluntary or charitable organisations if no food is stored on the premises.

Environmental health officers have sought a system of licensing for all food premises but the government believes that this is unnecessary bearing in mind the powers of enforcement provided by the Food Safety Act 1990.

1 Food Safety Act 1990, s 19.
2 Food Premises (Registration) Regulations 1991/2825.

Training of Food Handlers

12.1.4 The ministers also have the power to make regulations requiring persons engaged in food business and handling food to be trained.[1] Consultation on draft regulations began in mid-1991 and the regulations are expected to be in place by mid-1992. Food authorities are empowered to provide training courses in food hygiene, whether within or outside their area, for persons who are, or intend to become, involved in food businesses whether as proprietors or otherwise.[2] A food authority may contribute towards the expenses incurred by any other food authority or towards expenses incurred by any other person in providing such courses. Many food authorities already offer food hygiene training but the regulations mentioned above should improve such training by providing a national standard.

12.2.1 *Food Hygiene*

For the current law as to food handlers see **12.2.3** below.

1 Food Safety Act 1990, s 16 (1).
2 Food Safety Act 1990, s 23.

12.2 The Hygiene Regulations

12.2.1 The principal requirements of the food hygiene regulations (see **12.1.1** above) are summarised below. Food temperature requirements are discussed in Chapter 13.

General Requirements

12.2.2

(a) No food business may be carried on at any insanitary premises, place or stall or at any premises or place the condition, situation or construction of which is such that food is exposed to the risk of contamination.

Accomodation for the discharge, movement, storage or dispatch of any food must not be in such a state as to expose food to the risk of contamination.

(b) Articles or equipment with which food comes into contact must be kept clean and, with the exception of non-returnable containers, must be so constructed, be of such materials and be kept in such good order, repair or condition as to –

- enable them to be thoroughly cleaned;
- prevent, so far as is reasonably practicable, any matter being absorbed by them;
- prevent as far as is reasonably practicable, any risk of contamination of the food.

All containers intended for contact with food, whether or not such contact occurs, must as far as reasonably practicable, be kept free from contamination.

In determining whether any article or equipment is clean, regard must be had to the nature and packing of the food for which the article or equipment is required and to the use which is made of the article or equipment.

Similar requirements to those above apply to the condition of land, vehicles and stalls used in the course of a food business.

(c) With certain controlled exceptions, no food may be given out for preparation or packing by another person at any domestic premises other than those of the person carrying on the business. Domestic premises used for a food business are liable to inspection.[1]

1 Food Safety Act 1990, s 32 (1).

Food Handlers

12.2.3

(i) A person who engages in food handling must take all such steps as may be reasonably necessary to protect the food from risk of contamination, and in particular:

(a) shall not so place the food as to involve any risk of contamination;

(b) before offering any food for sale, must ensure that any food which is unfit for human consumption is kept apart from any other food;

(c) must not, in or about any forecourt or yard, place any food lower than 18 inches from the ground unless it is adequately protected from risk of contamination;

(d) must ensure where reasonably necessary that open food while exposed for sale or during sale or delivery is kept covered or is otherwise effectively screened from possible sources of contamination;

(e) must not keep in any food room any animal feed unless it is in a container of such material and so closed as to prevent the risk of contamination of the food in the room.

As to 'risk of contamination' of food, this means a risk which could render food injurious to health, eg open food over which persons may sneeze or which may be handled with dirty hands.[1]

(ii) A person handling food must maintain standards of personal cleanliness by –

(a) keeping as clean as may be reasonably practicable all parts of his person which are liable to come into contact with the food;

(b) keeping as clean as may be reasonably practicable all parts of his clothing or overclothing which are liable to come into contact with the food;

(c) keeping any open cut or abrasion on any exposed part of his person covered with a suitable waterproof dressing;

(d) refraining from spitting;

(e) refraining from the use of tobacco or any other smoking mixture or snuff while he is handling any open food or is in any food room in which there is open food. A person 'uses tobacco' if he holds a partly smoked but extinct cigarette between his lips. The purpose of the regulation is to prevent ash from falling on the food.[2]

(iii) A person handling open food, other than raw vegetables, intoxicating liquor or soft drinks, must, while so engaged, wear sufficiently clean and washable overclothing. Any person who carries open meat which is liable to come into contact with his neck or head must also wear a clean and washable neck and head covering.. The. foregoing does not apply to waiters in the catering business, persons carrying unskinned rabbits or hares or unplucked game or poultry; or to the transport of food by railway undertakers or other carriers using vehicles for the transport of food which is not ordinarily so used, provided that in all cases the person carrying the food takes all such other precautions as are reasonable and practicable to prevent the food from coming into contact with any exposed part of his person or with any clothing other than over-clothing.

(iv) A person handling food must not:

(a) carry any food in a container together with any article from which there is a risk of contamination of the food, or with any live animal or live poultry, without taking all such precautions as are reasonably practicable to avoid risk of contamination and, in particular, must not allow any live animal or live poultry to come into contact with any food;

(b) use for wrapping or containing any open food any paper or other wrapping material or container which is not clean or which is liable to contaminate the food, and must not allow any printed material, other than printed material designed exclusively for wrapping or containing food,[3] to come into contact with any food other than uncooked vegetables or unskinned rabbits or hares or unplucked game or poultry.

(c) Immediately a person engaged in the handling of food becomes aware that he is suffering from, or is the carrier of, typhoid, paratyphoid or any other salmonella infection or amoebic or bacillary dysentry or any staphylococcal infection likely to cause food poisoning, he must inform the person carrying on the food business and that person must immediately notify the appropriate medical officer of health. It is an offence for a person handling food to fail to notify his employer as required and an offence for the employer to fail to notify the medical officer of health.

1 *MacFisheries (Wholesale and Retail) Ltd v Coventry Corpn* [1957] 3 All ER 299, [1957] 1 WLR 1066.
2 *Pitt v Locke* (1960) 125 JP 93, DC.
3 Materials and Articles in Contact with Food Regulations 1987/1523.

Food Premises

12.2.4 The requirements applicable to food premises are:

(a) No fresh air intake or any ventilation pipe included in the soil drainage system of food premises may be situated in a food room. Every inlet into any such system situated in any such room must be trapped.

(b) No cistern for the supply of water to a food room may supply a sanitary convenience otherwise than through an efficient flushing cistern or some other flushing apparatus equally efficient and suitable for preventing contamination of water supplies.

(c) Every sanitary convenience situated in or used in connection with food premises must be kept clean and in efficient order and placed so that no offensive odours can penetrate into a food room. Any room or other place containing a sanitary convenience must be suitably lighted and ventilated and kept clean. No room which contains a sanitary convenience may be used as a food room, and no room which communicates directly with a room or other place containing a sanitary convenience may be used for the handling of open food.

There must be affixed and maintained in a prominent position near every sanitary convenience available for use by persons employed in the handling of food a clearly legible notice requesting users to wash their hands after using the convenience.

(d) Subject to any certificate of exemption given by the food authority[1] a clean, wholesome and constant supply of water must be provided at all food premises.

(e) All food premises must have suitable and sufficient wash- hand basins for the use of all persons engaged in the handling of food situated in a position conveniently accessible to such persons. At every such wash-hand basin there must be an adequate supply of hot and cold water or of hot water of a suitable controlled temperature or, in the case of food premises where no open food is handled, of cold water. Whether or not the wash-hand basins are conveniently accessible is a question of fact. It has been held[2] that a wash-hand basin in the living quarters of an employer which had to be approached by stairs was so convenient. These are believed to be somewhat perverse judgments which should not be relied upon. They clearly run against the objectives of the regulations and will doubtless be corrected when the regulations are revised.

Every such wash-hand basin must be kept clean and in good working condition. They must be supplied with soap or other suitable detergent, nail brushes and clean towels or other drying facilities. Failure to supply one or more of these items creates separate offences calling for separate informations.[3]

Washing facilities may not be used for any purpose other than securing the personal cleanliness of the user.

(f) All food premises must be supplied with a sufficient quantity of suitable bandages and dressings (including waterproof dressings) for first aid treatment of persons engaged in handling food. Such items must be readily accessible to such persons.

(g) Subject to any certificate of exemption provided by the food authority[1] all food premises where open food is handled must be provided with suitable accommodation for outdoor and other clothing or footwear not being worn during working hours. Such clothing may not be kept on or about the premises other than in such accommodation.

(h) All premises where open food is handled must be equipped with sinks or other washing facilities suitable and sufficient for any necessary washing of food and equipment used in the food business. Every such sink etc must be provided with an adequate supply of hot and cold water or of hot water of suitably controlled temperature. Cold water only is sufficient for washing fish, fruit or vegetables or for washing with a suitable bactericidal agent drinking vessels, or ice cream formers or servers. Every such sink etc must be kept clean and in good working order.

(i) Suitable and sufficient lighting must be provided in every food room.

(j) Except where humidity and temperature are controlled suitable and sufficient means of ventilation must be provided and maintained in every food room.

(k) No food room may be used as a sleeping place and no food room which communicates directly with a sleeping place may be used for the handling of open food.

(l) The walls, floors, doors, windows, ceiling, woodwork and all other parts of the structure of every food room must be kept clean and in such good order, repair and condition as to enable them to be effectively cleaned and prevent, so far as reasonably practicable, the entry of birds and any risk of infestation by rats, mice, insects or other pests. These requirements create two offences only, namely, (i) failure to keep the structure of every food room clean, and (ii) failure to keep the structure of every food room in good order, repair and condition so as to be effectively clean or so as to prevent any risk of infestation etc.[4]

(m) The layout of food premises must be such as to provide adequate space, suitably sited, for the purpose of the removal of waste from food and the separation of unfit food in compliance with **12.2.3** (b) above, and the storage of any such waste and unfit food prior to disposal.

No refuse or filth, whether solid or liquid, may be deposited or allowed to accumulate in a food room except so far as may be unavoidable for the proper carrying on of the trade or business.

1 Food Hygiene (General) Regulations 1970/1172, reg 28 (2).
2 *Storey v Manning* (1958) unreported; *Adams v Flook and Flook* [1987] BTLC 61.
3 *Berni Inns Ltd v Reynolds* [1981] CLY 2233.
4 *George v Kumar* (1982) 80 LGR 526.

12.3 Milk and Dairies[1]

Dairy Farms and Dairy Farmers

12.3.1 Any person carrying on the trade of dairy farmer and every dairy farm must be registered by the Minister of Agriculture, Fisheries and Food. Stringent hygiene provisions are laid down for the buildings and water supplies used in relation to dairy farming and there are special requirements as to the production of milk, its treatment, storage and handling. The minister may refuse to register a person carrying on the trade of dairy farmer or a dairy farm if the conditions and requirements of the regulations are not satisfied.

There are special provisions to protect milk against contamination or infection; for the inspection of diseased cows, the prevention of the sale of milk from diseased cows; the reporting of infection of persons working in the production or distribution of milk; and the cleanliness and suitability of vessels, receptacles and vehicles used for the storage and distribution of milk.

1 Milk and Dairies (General) Regulations 1959/277.

Milk Distributors

12.3.2 Every person carrying on the trade of a milk distributor must be registered with the food authority in whose district their premises are. Requirements as to the construction and cleanliness of buildings, equipment, receptacles are laid down.

Guidance

12.3.3 A booklet entitled *A Guide to Clean Milk Production* was published by the Ministry of Agriculture, Fisheries and Food in 1991.[1] It presents an excellent summary as to the legal requirements relating to milk production.

1 *A Guide to Clean Milk Production*: MAFF Publications, London, SE99 7TP – PB 0341.

12.4 Milk-based Drinks[1]

12.4.1 All milk-based drinks, that is liquid drinks (other than a fermented drink) being a mixture comprising by weight at least 85% milk with other specified ingredients, are subject to standards of hygiene for their production and handling.

1 Milk-based Drinks (Hygiene and Heat Treatment) Regulations 1983/1508.

Note: For hygiene requirements as to fresh meat and meat products see Chapter 14.

Chapter 13

Temperature Controls

13.1 Introduction

General Offences

13.1.1 If chilled, frozen or hot food is not kept at a safe temperature and, as a consequence becomes microbiologically unsafe so as to breach the food safety requirement it is an offence.[1] Certain direct requirements as to the temperatures at which chilled foods should be kept when delivered to caterers were imposed many years ago but in early 1991 there was a major extension of the law on food temperatures.

1 Food Safety Act 1990, s 8.

Food Hygiene Amendments

13.1.2 It would have been reasonable to expect such an important extension of the law to be the subject of new regulations. For reasons known only to themselves, the Department of Health introduced the new requirements by amendments to the food hygiene regulations.[1] The amendments were made because of growing awareness of the role temperature control can play in preventing foodborne illnesses and the increase in the range and variety of refrigerated foods on sale to the public. The consultation period and the drafting of the regulations was difficult because of the many problems associated with setting maximum or minimum temperatures for chilled foods and the difficulties involved in identifying foods which will support rapid multiplication of pathogenic micro-organisms if they are not so controlled. The 1990 regulations were quickly followed by amendments in 1991 and it seems likely that further changes will be necessary soon.

1 Food Hygiene (General) Regulations 1970/1172; Food Hygiene (Markets, Stalls and Delivery Vehicles) Regulations 1966/791; Food Hygiene (Docks, Carriers etc) Regulations 1960/1602 as amended by the Food Hygiene (Amendment) Regulations 1990/1431 and the Food Hygiene (Amendment) Regulations 1991/1343.

Quick-Frozen Foods

13.1.3 Regulations[1] have been made to implement EC requirements[2] on temperature and other conditions to be satisfied when quick-frozen foods are sold as such. The regulation do not apply to frozen foods unless they have been subject to a quick-freezing process as defined and are labelled as 'quick-frozen'. The logic of this is somewhat obscure.

1 Quick-frozen Foodstuffs Regulations 1990/2615.
2 Council Directive 89/108/EEC (OJ No L 400; 11.02.89).

Other Temperature Requirements

13.1.4 There are requirements for the keeping of specified foods above or below certain temperatures including milk, ice-cream, liquid egg, milk-based drinks, cream and meat. There are also requirements as to the temperatures below which perishable foods should be kept during international transport.[1] For a summary of these requirements see below.

1 International Carriage of Perishable Foodstuffs Act 1976; International Carriage of Perishable Foodstuffs Regulations 1985/1071.

13.2 Chilled and Hot Foods

Specified Temperatures and Commencement Dates

13.2.1 The regulations prescribe maximum temperatures of 8°C and 5°C (specified temperatures) for chilled foods and a minimum of 63°C for hot foods above or below which it is unlawful to keep the food. From 1 April 1991 all relevant foods were required to be kept at or below 8°C or at or above 63°C. This requirement includes vehicles over 7.5 tonnes gross weight but not transport in small vehicles under that weight making local deliveries. From 1 April 1992 the above temperature controls were extended to transport in small local delivery vehicles. From 1 April 1993 certain of the relevant foods must be maintained at or below 5°C or above 63°C. The lower temperature does not apply to food in small local delivery vehicles for no longer than 12 hours when the maximum allowable chill temperature is 8°C.

Foods to be Kept at or Below 8°C (Relevant Foods)

13.2.2 The relevant foods are:

(a) soft cheeses (whether whole or cut) which have been ripened by the action of moulds or other micro-organisms;
(b) cooked products (whether prepared ready for consumption or intended to be subject to reheating or further cooking prior to consumption comprising or containing:
  (i) meat,
  (ii) fish,
  (iii) eggs,
  (iv) substances used as a substitute for meat, fish or eggs,
  (v) cheese,
  (vi) cereals,
  (vii) pulses, or
  (viii) vegetables,
whether or not the food also includes other raw or partially-cooked ingredients;

(c) smoked or cured fish;
(d) smoked or cured meat which has been cut or sliced after smoking or curing;
(e) desserts, an ingredient of which is milk, or anything used as a substitute for milk and which have a pH value of 4.5 or more;
(f) prepared vegetable salads, including those containing fruit;
(g) cooked pies and pasties containing meat, fish or any substitute for meat or fish, or vegetables or cheese or any combination thereof, encased in pastry except (in each case) those into which nothing has been introduced after cooking and which are intended to be sold on the day of their production or the day following that day;
(h) cooked sausage rolls except those which are intended to be sold on the day of their production or the day following that day;
(i) uncooked or partly cooked pastry and dough products containing meat, fish or any substance used as a substitute for meat or fish;
(j) sandwiches, filled rolls and similar bread products containing –
  (i) meat,
  (ii) fish,
  (iii) eggs,
  (iv) substances used as a substitute for meat, fish or eggs,
  (v) soft cheeses as in (a) above,
  (vi) vegetables; or
(k) cakes containing cream or anything used as a substitute for cream.

Food to be Kept at or below 5°C after 1 April 1993

13.2.3 The following foods, which are included in the requirements above until 1 April 1993 must be kept at or below 5°C from that date:

(a) cheeses which have been cut up or otherwise separated from the whole cheese from which they were removed;
(b) food included in **13.2.2** (b) above which has been prepared for consumption without the necessity for further cooking or reheating;
(c) smoked or cured fish;
(d) smoked or cured meat which has been sliced or cut after smoking or curing; or
(e) sandwiches, filled rolls and similar bread products containing any of the foods mentioned in this paragraph, unless they are intended to be sold within 24 hours of their preparation.

Exempt Foods

13.2.4 In view of the difficulty of defining precisely the food included in the above requirements the regulations specifically exempt the following:

(a) bread, biscuits, cakes or pastry which are of a description specified in **13.2.2** above by reason only of the use of egg, milk or cheese as an ingredient in the products (other than filling in such products) before they are baked,
(b) filled cooked pastry products where –
  (i) they are relevant foods by reason only of the use of egg or egg and milk,
  (ii) the products are intended for sale within 24 hours of their production, and
  (iii) the products are not cut up after the conclusion of preparation;
(c) ice-cream to which the Ice-cream (Heat Treatment etc) Regulations 1959[1] apply;

(d) food which has been subjected to dehydration or other process the purpose of which is to prevent the growth of pathogenic micro-organisms at ambient temperatures;
(e) food contained in hermetically sealed containers and which has been preserved by a process of heating, the purpose of which is to prevent the growth of pathogenic micro-organisms at ambient temperatures;
(f) uncooked bacon, uncooked ham, dry pasta, dry pudding mixes or dry mixes for the preparation of beverages;
(g) chocolate or sugar confectionery; or
(h) milk which is not combined with other ingredients.

1 Ice-cream (Heat Treatment etc) Regulations 1959/734.

Modifications

13.2.5

(1) The temperature requirements do not apply to any relevant foods in any food room –
  (a) for a period of two hours from the conclusion of preparation, if it is intended –
    (i) to be sold within that period having been subjected during that period to any process of heating the purpose of which is to maintain the food at a temperature between the specified temperature and 63°C,
    (ii) to be re-heated to at least 63°C at any time prior to sale;
  After expiry of the period of two hours relevant food which has been subjected to heating as in (i) above must not be sold or offered for sale;
  (b) for a period of four hours from the conclusion of preparation if it is not to be subjected to any process of heating the purpose of which is to maintain the food at a temperature between the specified temperature and 63°C between the conclusion of preparation and the time of sale; or
  (c) displayed for a period not exceeding four hours on catering premises where such display is for the purpose either –
    (i) of indicating to prospective purchasers the nature of the food which is for sale for consumption on those premises, or
    (ii) of having it available for service to a purchaser for consumption on those premises,
  provided that in either case no more food is displayed than is reasonably necessary for that purpose.

(2) The preparation of any food in **13.2.2** (g) or (h) or a filled cooked pastry product the filling of which principally comprises a combination of egg and milk, but also contains other relevant food is treated as concluded when its baking is concluded.

(3) The temperature requirements do not apply to any food in **13.2.2** (g), (h) or (k) or filled cooked pastry products the filling of which principally comprises a combination of egg or milk but also contains other relevant food in any food room for a period of two hours from the conclusion of preparation. Where the process of bringing a food to the specified temperature is begun but not completed, before the expiry of the period of two hours, the period shall be treated as extended for so long as is reasonably necessary to bring the food to that temperature.

Special Provisions for Foods in Catering Premises

13.2.6 The temperature requirements do not apply to relevant food in any food room which is in any catering premises and is intended for consumption on those premises if the process of service of such food is intended to be commenced –

(a) within two hours of the conclusion of preparation if it is intended that the food be reheated to at least 63°C prior to the commencement of the process of service;

(b) within four hours of the conclusion of preparation of the food if the food is not to be subjected to any process of heating the purpose of which is to maintain the food at a temperature between the specified temperature and 63°C between the conclusion of preparation and the intended commencement of the process of service.

Variations of Temperature

13.2.7

(1) Relevant food which is in any food room –
   - (a) if it is not either at or below the specified temperature or at or above 63°C must be cooled to a temperature not exceeding the specified temperature or heated to at least 63°C as appropriate, without any avoidable delay after arrival or after any process of preparation has been completed;
   - (b) thereafter must be kept at or below the specified temperature or at or above 63°C until it is sold except where –
     - (i) it is being heated without any avoidable delay from the specified temperature to 63°C or is being cooled without any avoidable delay from 63°C to the specified temperature, or
     - (ii) having previously been cooled to a temperature at or below the specified temperature, or heated to a temperature at or above 63°C, it is in the course of being served on the catering premises.

(2) Where relevant food in any room is placed, without avoidable delay, in equipment which is capable of cooling it to, and keeping it at, the specified temperature, any delay in cooling it to that temperature may not be regarded as avoidable for the purposes of **13.2.7** (1) above merely because other equipment not in the food room could have cooled the food more quickly.

(3) The specified temperature may be exceeded for a period of not more than two hours by up to 2°C where the variation is due to –
   - (a) a process of preparation;
   - (b) defrosting of equipment;
   - (c) temporary breakdown of equipment;
   - (d) movement of food from one part of the premises to another; or
   - (e) some other unavoidable reason.

Mail Order etc Food

13.2.8 The temperature requirements do not apply to relevant food while it is in the course of conveyance by post or by a private or common carrier to the ultimate consumer. The Ministry of Agriculture, Fisheries and Food has produced a booklet on *The Microbiological Status of Some Mail Order Foods*[1]

1 *The Microbiological Status of Some Mail Order Foods*: MAFF Publications, London, SE99 7TT – ref: PB 0707.

Delivery Vehicles

13.2.9 The temperature requirements for delivery vehicles carrying relevant foods are:

(a) from 1 April 1991 delivery vehicles over 7.5 tonnes gross weight must keep and deliver the foods at or below 8°C or at or above 63°C;
(b) from 1 April 1992 delivery vehicles of 7.5 tonnes or under gross weight must keep and deliver the foods at or below 8°C or at or above 63°C;
(c) from 1 April 1993 delivery vehicles over 7.5 tonnes gross weight must keep and deliver the foods subject from that date to the lower specified temperature, at or below 5°C;
(d) from 1 April 1993 foods kept and delivered locally in small vehicles may be kept and delivered up to 8°C rather than 5°C if delivered within 12 hours. Otherwise the relevant food must be kept and delivered at or below 5°C.

Guidance on Enforcement

13.2.10 Guidance on the above temperature controls is available[1] and a code of practice has been issued for the guidance of food authorities on their enforcement.[2] They are of value to food businesses.

1 *Guidelines on the Food Hygiene (Amendment) Regulations 1990*: HMSO ISBN 0 11 321369 7.
2 Code of Practice No. 10: Enforcement of the Temperature Control Requirements of Food Hygiene Regulations (HMSO ISBN 0 11 321465 0).

13.3 Quick-frozen Foods

Definition and Description

13.3.1 The regulations[1] define a quick-frozen food product as one comprising food which has undergone a freezing process known as 'quick-freezing' whereby the zone of maximum crystallisation is crossed as rapidly as possible, depending on the type of product, and which is labelled for the purpose of sale to indicate that it has undergone that process, but does not include ice-cream or any other edible ice. The regulations do not, therefore, apply to a product, even though it may have undergone the quick-freezing process, unless it is labelled as such.

1 Quick-frozen Foodstuffs Regulations 1990/2615.

Labelling, Packaging and Temperature

13.3.2 Where a quick-frozen food is marked with the description 'quick-frozen' it must also be labelled with a sales name and:

(a) an indication of the date of minimum durability;
(b) an indication of the maximum period during which it is advisable to store it;
(c) an indication of one or other, or both of –
  (i) the temperature at which, and
  (ii) the equipment in which,
  it is advisable to store it;
(d) a reference allowing identification of the batch to which it belongs;
(e) a clear message of the type 'do not refreeze after defrosting'.

It is unlawful to sell to the ultimate consumer any quick-frozen foodstuff unless it has been packed by its manufacturer or packer in such prepackaging as is suitable to protect it from microbial and other forms of external contamination and against dehydration and it has remained in such prepackaging up to the time of sale.

Quick-freezing must result in the temperature of the food after thermal stabilisation being −18°C or colder and it must be maintained, subject to permitted exceptions, at that temperature or below. There are special requirements as to the equipment to be used by the manufacturer, storer, transporter, local distributor and retailer.

13.4 International Carriage of Perishable Foodstuffs

13.4.1 The Act[1] is the United Kingdom's accession to the Agreement on the International Carriage of Perishable Foodstuffs (APT) which was concluded in Geneva on 1 September 1970.[2] The regulations[1] give effect to the APT, which is frequently amended, by laying down the maximum temperatures at which certain foods may be carried on loading, during carriage and on unloading.

1 International Carriage of Perishable Foodstuffs Act 1976; International Carriage of Perishable Foodstuffs Regulations 1985/1071.
2 Cmnd 6441.

13.5 Other Temperature Requirements

13.5.1 Regulations made under or continued in force by the Food Safety Act 1990 impose a number of special temperature requirements in addition to those relating to heat treatment such as pasteurisation, sterilisation and ultra high temperature. They are:

(a) *Milk*: Dairy farmers are required to cool milk immediately after production, unless it is to be heat treated on the premises, to either a maximum of 10°C or, if the temperature of the water supply available for cooling is 7°C or above, to a maximum of 3°C. Distributors who receive milk which has not been previously cooled to a maximum of 10°C must, subject to exceptions, cool it to that temperature.[1]

(b) *Ice-cream*: may not be offered for sale unless it has been kept at a temperature not exceeding 28°F since it was frozen, or if its temperature has risen above 28°F at any time since it was frozen, it has again been subjected to the treatment (if any) to which it is required to be subjected by the regulations and, after treatment has been kept at a temperature not exceeding 28°F.[2]

(c) *Liquid egg*: which has not been heat treated as required by the regulations but which is removed from shell on the premises where the food is prepared must either be used forthwith or kept at a temperature not exceeding 10°C and used within 24 hours.[3]

(d) *Meat and meat products*: Fresh meat and meat products must be transported in sealed means of transport, designed and equipped in such manner so that the following temperatures for meat are maintained during the course of transport:
   (i) for carcases and cuts (excluding rabbit meat, hare meat and poultry meat) – max + 7°C (chilled),

(ii) for offals (excluding rabbit offals, hare offals and poultry offals) – max – 12°C (frozen),

(iii) for rabbit meat, hare meat, poultry meat, rabbit offals, hare offals and poultry offals – max + 4°C,

(iv) for meat products – temperatures as specified on the label when appropriate.[4]

1 Milk and Dairies (General) Regulations 1959/277.
2 Ice-cream (Heat Treatment etc) regulations 1959/734.
3 Liquid Egg (Pasteurisation) Regulations 1963/1503.
4 Imported Food Regulations 1984/1918.

Chapter 14

Imported Food and Health Problems Arising from Food

14.1 Background

Historical Control of Imported Food

14.1.1 For the past 50 years food imported into the United Kingdom has been controlled by national regulations[1] culminating in regulations of 1984[2] which, by amendment, have incorporated relevant EC requirements up to 1992. The purpose of these regulations has been to ensure that imported food is safe for consumption and, if it is not, that it can be stopped at the point of importation before release into the retail market. To that end the primary agencies of enforcement have been port health authorities working with HM Customs and Excise. By any objective assessment the work of these agencies has been successful, for standards of hygiene and safety in imported food have rarely been a cause of public concern.

Since the United Kingdom's accession to the Treaty of Rome in 1972, the imported food regulations of the day have been amended to implement Community requirements. As is the fashion with Community law the subject matter of the relevant legislative instruments is not confined to a general objective. Instruments on public health matters sometimes include compositional requirements; instruments concerning food imported into the Community also provide for internal matters and so on. This has resulted in EC measures being split between a number of United Kingdom regulations.

The Food Safety Act 1990[3] has greatly strengthened the powers of the ministers and enforcement authorities to prevent the importation of unsafe food.

1 Public Health (Imported Food) Regulations 1937 and 1948; Imported Food Regulations 1963.
2 Imported Food Regulations 1984/1918 as amended.
3 Food Safety Act 1990, ss 16 (1) (f), 17 (1) and 18 (1) (c).

National and EC Considerations

14.1.2 The law on imported foods has, until now, always been concerned with food imported into the United Kingdom. The definition of 'import' (see **4.3.1**) confines the term to foodstuffs brought into the United Kingdom and the Imported Food Regulations 1984 apply similarly. However, new and impending EC measures on meat, meat products, milk, eggs and a variety of other product sectors will apply to foodstuffs brought into the Community. Accordingly, the law discussed in this chapter is transitory but efforts have been made to anticipate the changes which will occur from 1 January 1993 when the European Single Market comes into operation. Inevitably, this chapter is somewhat speculative.

14.2 General Regulations

Requirements Applicable to all Imported Food[1]

14.2.1 It is an offence to import food intended for human consumption –

(a) which has been rendered injurious to health by the addition of any article or substance to it; by using any article or substance as an ingredient in the preparation of the food; by abstracting any constituent from it and by subjecting it to any other process or treatment (see **5.1.2**);
(b) which has been examined by a competent authority and found at the time of examination not to be fit for human consumption;
(c) which is otherwise unfit for human consumption or is unsound or unwholesome;
(d) in the preparation of which any such food as aforesaid has been used.

These offences include provisions in ss 7 and 8 of the Food Safety Act 1990 (see Chapter 5), but additionally include the offences of 'unsoundness' and 'unwholesomeness' which are unique to imported food law. Whether or not food is unsound or unwholesome is a question of fact but note that food which is unfit, unsound or unwholesome but which will be processed to remove the unfitness etc is not necessarily contrary to these requirements.[2] The terms 'unsoundness' and 'unwholesomeness' may be contrasted with the food safety requirement which is discussed in Chapter 10.

Although duplicating some elements of the offences created by s 8 of the Food Safety Act 1990, this regulation does not require proof of 'sale' etc for an offence to have been committed. Its purpose is to prevent the importation of unsafe food and it is merely necessary to prove importation with an intention that the food should be sold for human consumption. However, successful control of unsafe imported food does not depend upon a prosecution because the food safety requirement is applied indirectly to imported food by virtue of the powers of authorised officers (see **14.2.3** below).

1 Imported Food Regulations 1984/1918, reg 6.
2 *R v Archer, ex p Barrow, Lane & Ballard Ltd* (1983) 147 JP 503, CA.

Special Defence

14.2.2 It is a defence to a charge under the above provisions for the defendant to prove that at the time when the food was imported he did not know, and could not with reasonable diligence have ascertained, that it was unfit for human consumption or unsound or unwholesome. For a discussion of the general defence of 'all reasonable precautions and all due diligence' in the Food Safety Act 1990[1] (applied to this regulation)[2] see chapter 19.

1 Food Safety Act 1990, s 21.
2 Imported Food Regulations 1984/1918, reg 2A.

Examination of Imported Food[1]

14.2.3 Authorised officers may, at all reasonable times, examine any imported food intended for sale for human consumption. Where, on such examination, it appears to the officer that such food contravenes the provisions discussed above he may by notice in writing to the importer and/or any other person in possession of the food require that until the food has been dealt with by a justice of the peace it shall not, without the consent of the officer, be moved from the place of examination or any other place specified in the notice. If it appears to the officer that the food contravenes the above provisions it may be treated as having failed the food safety requirement and the officer may exercise his powers of seizure etc.[2]

1 Imported Food Regulations 1984/1918, reg 7.
2 Food Safety Act 1990, s 9 (see also **10.2**).

The Notice

14.2.4 The notice may be deemed to have been issued under s 9 of the Food Safety Act 1990 and the prescribed form of notice[1] and method of service[2] is appropriate. The authorised officer should also have regard to the appropriate codes of practice.[3]

The notice must specify the grounds on which it is based and if the food is subject to any EC directives concerning health problems affecting intra-Community trade in meat (see **14.4** below) the notice must inform the addressee of his right to serve a counter-notice.

After service of a notice the authorised officer must notify the minister of the notice giving full details of the food and the reason for its rejection or condemnation.

1 Detention of Food (Prescribed Forms) Regulations 1990/2614).
2 Food Safety Act 1990, s 50.
3 Code of Practice No 4: Inspection, Detention and Seizure of Suspect Food (HMSO ISBN 0 11 321350 6).

Counter Notices

14.2.5 Where the food about which a notice has been issued is subject to one of the EC meat directives the recipient of the notice may serve a counter notice in writing on the authorised officer requiring him, before the food is dealt with by a justice of the peace, to obtain the opinion of an independent veterinary expert, to be nominated by the appropriate minister, as to the condition of the food and the validity of any matter specified in the notice as a ground on which it is based. Upon receipt of such a notice the authorised officer must inform the appropriate minister of its contents and request the nomination of the independent veterinary expert.

14.2.6 *The minister*, on being notified of a counter-notice must determine whether the food is subject to any of the EC meat directives. If it is not, then he must inform the person serving the counter-notice and the food must be dealt with by a justice of the peace. If it is, then the minister must consult the EC Commission as to the appointment of the independent veterinary expert who must not be a national of the exporting member state. That expert must then examine the food and give his written opinion thereon to the authorised officer who, within seven days of receipt, must make it available to the person who issued the counter-notice. If the opinion supports the officer's notice then the food must be dealt with by a justice of the peace. If the opinion does not support the notice then the authorised officer may rescind the original notice.

Special Examinations[1]

14.2.7 If the authorised officer is of the opinion that a special procedure is necessary for the examination of the food, or if the importer should request such examination, the importer must provide all such facilities as the authorised officer may reasonably require for the examination. An authorised officer may by notice in writing to the importer or any other person in possession of food, restrict the removal or delivery of the food during any period not exceeding six working days which may be reasonably required for the examination of the food.

1 Imported Food Regulations 1984/1918, reg 9.

Appeals[1]

14.2.8 Any person who has received a notice issued by an authorised officer may appeal against that notice to a magistrates' court and the court may direct that the notice be withdrawn or that such shorter period be fixed as appears reasonable.

1 Imported Food Regulations 1984/1918, reg 9.

Officers of Customs and Excise[1]

14.2.9 These officers play an important role in the control of imported food and work in close liaison with authorised officers of port health authorities. Where the duties of an officer of Customs and Excise in relation to imported food have not been wholly discharged, no examination of it may be made nor sample from it procured without his consent but he may not unreasonably withhold such consent. An authorised officer of a port health authority may request an officer of Customs and Excise to prohibit the removal of imported food which has not been cleared from customs charge until it has been examined by the authorised officer. Arrangements for such consents and examinations are prescribed by the regulations.

1 Imported Food Regulations 1984/1918, regs 4, 5, 10 and 11.

14.3 Imported Meat

Introduction

14.3.1 The Imported Food Regulations 1984 contain detailed requirements as to health marks and health certificates for imported meat, the disposal of meat imported contrary to the regulations and related matters. Recent and impending EC Directives on health problems relating to meat and meat products of different classes (see **14.4** below) will require substantial amendment of the regulations or their replacement. As indicated in **14.1.1** above, these directives include provisions not directly related to the safety of intra-Community trade in meat and meat products and it is not yet clear how they will be dealt with by the Ministry of Agriculture, Fisheries and Food.

The 1984 regulations take cognisance of EC measures dating from 1964 to 1977 but those measures have been or will shortly be replaced by more detailed ones dealing not only with the safety of imported meat but also with the hygiene of premises and, in some cases, with the quality of the meat or meat products.

Meat, Meat Products and Exemptions

14.3.2 The Imported Food Regulations 1984 apply to fresh meat and meat products which are defined as including any product prepared wholly or partly from fresh meat (which has undergone treatment to ensure a certain degree of preservation). This does not include fresh meat and vitamin concentrates containing meat or pharmaceutical products containing meat and gelatine.

Health Marks

14.3.3 It is an offence[1] to import for sale for human consumption any fresh meat, meat product or bulk lard unless it bears a health mark which is placed so as to be legible and clearly visible and conforms to prescribed requirements[2] which are recognised by the ministers. It is also an offence[1] to import for sale for human consumption any fresh meat or meat product unless it has been wrapped and packed, and is transported, in accordance with prescribed requirements.[3]

1 Imported Food Regulations 1984/1918, reg 13.
2 Imported Food Regulations 1984/1918, Sch 3.
3 Imported Food Regulations 1984/1918, Sch 4.

Health Certificates

14.3.4 It is an offence[1] to import for sale for human consumption any fresh meat which is derived from domestic bovine animals (including buffalo), swine, sheep, goats, solipeds or poultry, or any meat product (other than certain extracts, crushed bones, offal etc) in which any fresh meat is used unless it is accompanied by a health certificate which is valid and in a prescribed form.[2] Provision is made for lost, damaged or defective health marks and certificates.[3]

1 Imported Food Regulations 1984/1918, reg 14.
2 Imported Food Regulations 1984/1918, Schs 6, 7, 8, 9 and 10.
3 Imported Food Regulations 1984/1918, reg 15.

Disposal of Meat Imported Contrary to the Regulations[1]

14.3.5 The regulations make detailed provisions similar to those discussed at **14.2** above for the issue of notices by authorised officers, the issue of counter-notices by the importer or person in possession of the meat etc; the appointment of independent veterinary experts, submission of meat etc to a justice of the peace and the destruction or disposal of unfit meat etc.

1 Imported Food Regulations 1984/1918, reg 16.

14.4 EC Directives on Health Problems

The Scope

14.4.1 There is a substantial raft of Community law on health problems affecting meat, fish and eggs and products derived from all of these foods. The directives are complex and deal with the health of animals, the slaughter and cutting up of them and the treatment of the final products. These measures have an unhappy history resulting from confused proposals, over-hasty enactment and bad drafting. In January 1992 the proposed programme of measures is still not complete leaving precious little time for member states to implement them in time for the Single Market on 1 January 1993.

Current EC legislation on health problems arising from these foods, is listed in Appendix B. In addition, the Directive on Meat Products was adopted in December 1991 and publication in the *Official Journal of the EC* is awaited.

The Red Meat Directive[1]

14.4.2 This Directive re-enacts with substantial modifications Community law on intra-Community trade in fresh meat (except poultry meat) which has been in force ever since the United Kingdom joined the Common Market. A major departure from previous policy, however, is that it is now extended to national trade. Member states are required to implement the Directive by 1 January 1993. In the United Kingdom implementation will require the amendment and/or replacement of a number of relevant regulations.[2]

The Directive applies to meat intended for human consumption from domestic animals, being bovine animals (including the species *Bubalus bubalis* and *Bison bison*), swine, sheep and goats and domestic solipeds. The Directive does not apply to the cutting and storage of fresh meat performed in retail shops or in premises adjacent to sale points, where the cutting and storage are performed solely for the purpose of supplying the consumer directly on the spot.

The principal requirements of the Directive include:

Carcases, half carcases or half carcases cut into more than three wholesale cuts and quarters must:

(a) have been obtained at a licensed and approved slaughterhouse;
(b) come from an animal which has been inspected and passed as fit for slaughter;
(c) have been treated under satsifactory hygiene conditions as prescribed;
(d) have been inspected post-mortem by an official veterinarian (OV) with satisfactory results as prescribed;
(e) bear a health mark as prescribed;
(f) be accompanied during transportation by the health certificate issued by the OV at the time of loading and, from 30 June 1993, by a prescribed commercial document authorised by the OV;
(g) be stored after post-mortem inspection under satisfactory hygiene conditions in approved establishments;
(h) be transported under satisfactory hygiene conditions as prescribed.

Cuts or pieces smaller than those referred to above must:

(a) be boned or cut in a cutting plant satisfying prescribed conditions;
(b) be boned or cut and obtained as laid down and come from fresh meat which complies with prescribed conditions or fresh meat imported from third countries in accordance with Community law;[3]
(c) be stored under prescribed conditions;
(d) have been checked by the OV as prescribed;
(e) meet wrapping and packing requirements laid down.

Offal must:

(a) come from an approved slaughterhouse or an approved cutting plant and satisfy prescribed conditions;
(b) not be sliced unless they are livers from bovine animals and then only in an approved cutting plant. Derogations for the slicing of other livers may be given.

Cold stored fresh meat must:

(a) not have undergone any other handling except in connection with storage;
(b) satisfy prescribed conditions;
(c) be accompanied during transportation to the place of destination by commercial documents or certificates approved by the OV;
(d) if it has been cold stored in a third country approved by Community law[4] and under customs control, must not have undergone any other handling except in connection with storage;
(e) satisfy prescribed conditions;
(f) satisfy specific guarantees concerning checking and certification of compliance with storage and transport conditions;
(g) be accompanied by a prescribed certificate.

Slaughterhouses must be subject to veterinary registration, satisfy stringent conditions and be inspected by competent authorities and/or Community veterinary experts.

The OV must declare meat as unfit for human consumption if it is affected by one of a number of specified diseases, or contains residues of authorised substances, medicinal products or other substances which render it harmful to human health, or is otherwise contaminated.

Meat which is declared unfit for human consumption must be clearly distinguished and dealt with in a prescribed manner.

1 Council Directive 91/497/EEC (OJ No L 268; 24.09.91) amending and consolidating Directive 64/433/EEC.
2 Slaughterhouse (Hygiene) Regulations 1977/1805; Imported Food Regulations 1984/1918; Meat (Sterilisation and Staining) Regulations 1982/1018; Meat Inspection Regulations 1987/2236; Fresh Meat Export (Hygiene and Inspection) Regulations 1987/2237; Bovine Offal (Prohibition) Regulations 1989/2061.
3 Council Directive 90/675/EEC (OJ No L 373; 31.12.90).
4 Council Directive 72/462/EEC (OJ No L 302; 31.12.72).

Poultry Meat

14.4.3 *Poultry meat* is subject to similar but much earlier requirements.[1] This Directive is under review and will be consolidated and modified in due course.

1 Council Directive 71/118/EEC (OJ No L 055; 08.03.71) as amended, and in particular by Council Directive 88/657/EEC (OJ No L 382; 31.12.88).

Sheep and Goat Meat

14.4.4 There are EC measures on sheep and goats and products derived from them.[1] They concern health conditions governing intra-Community trade in ovine and caprine animals.

1 Council Directive 91/68/EEC (OJ No L 046; 19.02.91).

Rabbit Meat and Farmed Game Meat[1]

14.4.5 EC requirements for meat from domestic rabbits and all wild land mammals and wild birds are similar to those relating to poultry meat (see **14.4.3** above) but contain special requirements for dealing with health problems peculiar to meat obtained from such animals and birds.

1 Council Directive 91/495/EEC (OJ No L 268; 24.9.91).

Fishery Products[1]

14.4.6 All seawater or freshwater animals or parts thereof, including their roes, but excluding aquatic mammals, frogs and aquatic animals covered by other EC legislation (fishery products) and all fishery products born and raised in controlled conditions until placed on the market as a foodstuff (aquaculture products), must comply with new EC requirements as to the hygiene of ships, containers and premises with which they come into contact. Certain poisonous fish and fishery products containing certain biotoxins are prohibited. There are separate requirements as to bivalve molluscs.[2]

1 Council Directive 91/493/EEC (OJ No L 268; 24.09.91).
2 Council Directive 91/492/EEC (OJ No L 268; 24.09.91).

Egg Products[1]

14.4.7 Egg products (that is products obtained from eggs, their various components or mixtures thereof; after removal of the shell and membranes, intended for human consumption; which may be partially supplemented by other foodstuffs or additives; and may be liquid, concentrated, dried, crystallised, frozen, quick-frozen or coagulated) are subject to EC requirements which lay down standards of hygiene for the products themselves and the establishments in which they are made. The egg products subject to control are those from hens, ducks, geese, turkeys, guinea fowls and quails but do not include a mixture of eggs of different species.

1 Council Directive 89/437/EEC (OJ No L 137; 02.06.88).

Milk[1]

14.4.8 The hygienic handling and production of heat treated milk is laid down by three EC instruments. Standards of hygiene for milk tankers, premises, installations and working equipment are prescribed and milk treatment establishments must be approved.

1 Council Directive 85/397/EEC (OJ No L 226; 24.08.85); Council Regulation (EEC) No 382/89 (OJ No L 044; 16.02.89); Commission Directive (EEC) No 89/362 (OJ No L 156; 08.06.89).

14.5 Slaughterhouses

Introduction

14.5.1 Control of slaughterhouses, once a part of primary food law, is now achieved by the Slaughterhouses Act 1974 and regulations made thereunder.[1] They are concerned both with the welfare of animals going to slaughter and the hygiene of meat derived from them. The Act requires that slaughterhouses and knackers' yards (premises used for slaughtering, flaying or cutting up animals whose flesh is not intended for human consumption) be licensed. The provisions relating to hygiene are very briefly summarised below.

1 Slaughterhouse (Hygiene) Regulations 1977/1805 as amended.

Slaughterhouse Hygiene

14.5.2 The regulations[1] require that:

(a) slaughterhouses be arranged so as to facilitate the efficient performance of meat inspection; to permit clean operations to be carried out, adequately separated from those operations liable to give rise to contamination or pollution; and permit the functioning of all operations under hygienic conditions;
(b) accommodation must contain suitable and sufficient lairage, space for hanging meat, refrigerated storage, space for gutting, separate accomodation for meat requiring further examination or rejected meat, separate accomodation for the preparation of tripe and casings and storage of hides, storage for clothing and provision for cleaning vehicles;
(c) there must be artificial lighting of not less than 220 lux which does not distort the colour of meat;
(d) there must be proper ventilating apparatus;
(e) premises must be lined with smooth, durable, impervious and washable materials;
(f) there must be satisfactory drainage with traps for solids;
(g) there must be a sufficient supply of clean and wholesome hot and cold water;
(h) sanitary conveniences must be flushable and not in communication with the slaughterhall;
(i) there must be adequate washing facilities for staff;
(j) equipment must be durable and impervious to corrosion and capable of being kept clean;
(k) overhead rails, appropriate receptacles and other equipment must be provided as required;
(l) hygienic practices as to the admission of animals and carcases are laid down.

1 Slaughterhouses (Hygiene) Regulations 1977/1805 as amended.

14.6 Prohibition of Commercial Practices and Importation

Powers[1]

14.6.1 The Food Safety Act 1990 has greatly increased the powers of the ministers to prohibit, or to permit subject to conditions, commercial practices in relation to food, food sources and contact materials. Similarly, powers under the Food and Environment Protection Act 1985 (see Chapter 11) enable emergency prohibitions to be made. These powers have been exercised during 1991 on a number of occasions where the ministers have believed that there was a risk to human health.

1 Food Safety Act 1990, ss 16 (1) (f), 17 (1) and 18 (1) (c).

Imports from South America[1]

14.6.2 In early 1991 the ministers were concerned about the risk of cholera entering the United Kingdom from Peru by way of certain food products from that country. Regulations[2] forbade the importation of the suspected foods from that country. Later, concern spread to foods from Colombia and Ecuador in addition to Peru, so those regulations were revoked and replaced by new regulations which made it an offence to import any bivalve molluscs from Peru or any fish product obtained in the course of artisanal fishing, or fruit or vegetables from Peru, unless they are accompanied by a numbered and dated certificate from an officer of El Centro de Certificaciones Pesqueras in El Callao (CERPER). Importation of all fish products and bivalve molluscs from Ecuador and Colombia was prohibited unless they were accompanied by a similar certificate from the proper authority in those countries. The export of such products to other member states of the EC was also prohibited subject to certain conditions.

1 Imported Food and Feedingstuffs (Safeguards against Cholera) Regulations 1991/2486.
2 Imported Food (Peruvian Foodstuffs) Regulations 1991/370.

Hormonal Substances[1]

14.6.3 The administration of hormonal substances to animals is prohibited and special enforcement powers are conferred on authorised officers to enable them to give notices requiring that no commercial operations be carried out with respect to an animal and to restrict the movement of such animals. The authorised officer may also require the slaughter of an animal where a prohibited substance is proved to be present in it.

1 Hormonal Substances (Food Sources) (Animals) Regulations 1991/1593 as read with the Medicines (Stilbenes and Thyrostatic Substances) Regulations 1982/626, reg 3 and the Medicines (Hormone Growth Promoters) (Prohibition of Use) Regulations 1988/705, reg 3.

Bivalve Molluscs, Crabs etc

14.6.4 During 1991 there was concern about the danger of paralytic shellfish poisoning from certain bivalve molluscs and crabs caught in certain waters around the northern coasts of Scotland. Orders were made under the Food and Environment Protection Act 1985 to prohibit the removal of such shellfish from designated areas.[1] As soon as the danger was thought to have passed the orders were amended or revoked.

1 See for example the Food Protection (Emergency Prohibitions) (Paralytic Shellfish Poisoning) (No. 10) Order 1991/1958 (S 164).

Bovine Spongiform Encephalopathy (BSE)

14.6.5 Arising from public concern about the possible spread of BSE the government made regulations[1] prohibiting, subject to certain exemptions, the sale or use in food of any specified bovine offal and requiring that such specified bovine offal be sterilised or stained. The removal of such offal from slaughterhouses was prohibited, as was any freezing of it.

Action was also taken under powers conferred on the ministers by the Animal Health Act 1981[2] by way of an order[3] requiring the compulsory notification of BSE in bovine animals, the testing of such animals for BSE, the prohibition of certain feeding stuffs, the prohibition of the sale or supply of milk from affected or suspected animals and related matters.

1 Bovine Offal (Prohibition) Regulations 1989/2061 as amended.
2 Animal Health Act 1981, ss 1, 7 (1), 8 (1), 15 (4), 32 (2), 35 (1) and (1A), 36 (5) (a), 72, 83 (2), 87 (2) and (5) and 88 (2).
3 Bovine Spongiform Encephalopathy Order 1991/2246.

Chapter 15

Pesticides and Veterinary Residues

15.1 Introduction

15.1.1 The potentially harmful effect on human health from the manufacture, storage and use of pesticides and residues of pesticides in or on foodstuffs was recognised some 30 years ago. Initially it was believed that the general provisions of the Food and Drugs Act 1955, and later the Food Act 1984, would be sufficient to protect foodstuffs by virtue of the offences contained therein. The manufacture, storage and use of pesticides was subject to voluntary control by the pesticides industry in cooperation with the Ministry of Agriculture, Fisheries and Food. However, by 1985 it was apparent that there was need for a statutory framework of controls and this was enacted in Part III of the Food and Environment Protection Act 1985. That Part of the Act was essentially an enabling measure to empower the ministers to make regulations to establish a comprehensive control system backed, as appropriate, by statutory codes of practice.

The necessary regulations[1] were made in 1985, 1986 and 1988.

1 Control of Pesticides (Advisory Committee of Pesticides) Order 1985/1516; Control of Pesticides (Advisory Committee of Pesticides) (Terms of Office) Order 1985/1517; Control of Pesticides Regulations 1986/1510; Pesticides (Maximum Residue Levels in Food) Regulations 1988.

Advisory Committee on Pesticides

15.1.2 This committee was established on 31 October 1985 with a remit to advise the ministers on any matters relating to the control of pesticides in furthering the general purposes of Part III of the Food and Environment Protection Act 1985. Following consultation with the Committee the ministers made the Control of Pesticides Regulations 1986 which came into force on 1 July 1987. The transition from a voluntary control system to a statutory one was achieved quickly and efficiently.

Scope of the Regulations

15.1.3 The regulations apply to pesticides and any substance, preparation or organism prepared or used for the following purposes as if it were a pesticide:

(a) protecting plants or wood or other plant products from harmful organisms;

(b) regulating the growth of plants;
(c) giving protection against harmful creatures;
(d) rendering such creatures harmless;
(e) controlling organisms with harmful or unwanted effects on water systems, buildings or other structures, or on manufactured products;
(f) protecting animals against ectoparasites.

The regulations do not apply to substances controlled by the Medicines Act 1968, the Agriculture Act 1970 (Part IV), the Food Safety Act 1990 and the Cosmetic Products (Safety) Regulations 1984; substances used for disinfecting, bleaching or sterilising other substances (including water) other than soils, compost and other growing media; substances used in laboratories; certain substances used for the stimulation of growth of plants etc; and pesticides used in adhesives and for manufacturing or industrial purposes.

Prohibitions

15.1.4 It is an offence to:

(i) advertise a pesticide unless the ministers have given provisional or full approval to that pesticide and a consent to such advertising;
(ii) fail to obey any conditions attached to the approval of a pesticide by the ministers as to advertising;
(iii) sell, supply or store a pesticide unless the Ministers have given a provisional or full approval in relation to that pesticide and a consent as to such sale;
(iv) fail to obey any conditions relating to the approval or consent;
(v) use a pesticide unless the ministers have given an approval to that pesticide and a consent to such use;
(vi) fail to obey additional conditions as to pesticides applied from an aircraft in flight.

Approvals

15.1.5 The ministers may give their approval in relation to a pesticide in the form of an experimental permit to enable testing and development to be carried out with a view to providing the ministers with data as to safety etc; a provisional approval for a stipulated period with a view to satisfying the ministers' outstanding data requirements; or a full aproval for an unstipulated period.

An approval may be given subject to conditions imposed when or after it is given and the ministers may, at any time, review, revoke or suspend approval, or amend any conditions of an approval.

Consents

15.1.6 The ministers may by notice in the London and Edinburgh Gazettes give their consent to the advertisement of pesticides subject to prescribed basic conditions; the sale, supply or storage of pesticides subject to prescribed basic conditions; and the use of pesticides subject to prescribed basic conditions including additional conditions for aircraft application.

Seizure or Disposal of Pesticides

15.1.7 If there has been a breach, in relation to any pesticide, of any of the prohibitions given above, or of a condition of an approval or consent the ministers have the power to:

(a) seize or dispose of the pesticide or require that some other person shall dispose of it;
(b) seize or dispose of anything treated with the pesticide or require that some other person shall dispose of any such thing;
(c) direct some other person to take such remedial action as appears to the ministers to be necessary as a result of that contravention.

Release of Information to the Public

15.1.8 Following provisional or full approval of a pesticide the ministers may, at the request of any person, make available to him for inspection, on such conditions as they may determine, an evaluation of the pesticide in relation to which that approval was given or any conditions which were amended. If a person satisfies the ministers that an evaluation made available to him for inspection gives insufficient information for his purposes, the ministers may make available for his inspection at such times and on such conditions as may be determined by the ministers the study reports or other data supplied in support of an application for approval. Such evaluations or study reports may not be published without the consent of the ministers and fees may be payable for copies of such documents.

Basic Conditions for Advertising

15.1.9 The regulations lay down the conditions subject to which the ministers may consent to the advertising of a pesticide. Such advertisement must relate only to such uses of a pesticide as are permitted by the approval given in relation to that pesticide and any printed or pictorial advertisement however transmitted must include –

(a) a statement of the active ingredient of each pesticide mentioned in the advertisement;
(b) such general warning as the ministers may require;
(c) where required by a condition of approval given in relation to a pesticide mentioned in the advertisement, a statement of any special degree of risk to human beings, creatures, plants or the environment.

Basic Conditions for Sale, Supply and Storage

15.1.10 The regulations lay down the conditions subject to which consents to the sale, supply and storage of pesticides may be given. They are –

(a) persons who sell, supply or store pesticides must take all reasonable precautions to protect the health of human beings, creatures and plants and to safeguard the environment and be competent for the duties they are called upon to perform;
(b) persons who sell or supply pesticides for agricultural use or store such pesticides must either obtain a certificate of competence recognised by the ministers or sell or supply the pesticides under the supervision of another person who holds such a certificate;

(c) employers must ensure that any person in their employ who sells, supplies or stores pesticides is given instruction and guidance to enable him to achieve the standard of competence recognised by the ministers and to comply with the regulations.

Basic Conditions for Consent to Use

15.1.11 The regulations lay down the conditions subject to which consent to the use of pesticides may be given. They are –

(a) persons using pesticides must take all reasonable precautions to protect the health of human beings, creatures and plants, to safeguard the environment and to avoid pollution of water;
(b) no two or more pesticides may be mixed unless done in accordance with approvals;
(c) no pesticide may be used in conjunction with an adjuvant except in accordance with an approval;
(d) persons may not use pesticides in the course of business unless they have received adequate instruction and guidance in the safe, efficient and humane use of pesticides and are competent for their duties;
(e) persons may not use pesticides approved for agricultural use in the course of commercial services unless they have obtained certificates of competence recognised by the ministers or work under the direction and supervision of a person who holds such a certificate;
(f) no person who was born later than 31 December 1964 may use a pesticide approved for agricultural use unless he has obtained a certificate of competence recognised by the ministers, or he uses the pesticide under the direct and personal supervision of a person who holds such a certificate;
(g) every employer has a duty to ensure that persons in their employment who may be required to use a pesticide are provided with such instruction and guidance as is necessary to enable those persons to achieve the standard of competence recognised by the ministers and otherwise to comply with the regulations.

Pesticides Applied from Aircraft

15.1.12 There are stringent conditions as to consultation with and the giving of notices to the Nature Conservancy Council, the appropriate water authority, the Chief Environmental Health Officer for the area concerned, the occupants of nearby buildings, hospitals, schools and other institutions in the area and local beekeepers at specified intervals before the commencement of aerial application of pesticides. There are also restrictions imposed on the pilots of the aircraft.

Codes of Practice

15.1.13 The Act[1] provides for the issue of codes of practice for the purpose of providing practical guidance in respect of any provisions of the Act or of regulations. The codes issued to date are –

(a) The Code of Practice for the Safe Use of Pesticides on Farms and Holdings (1990);

(b) The Code of Practice for Suppliers of Pesticides to Agriculture, Horticulture and Forestry (1990).

1 Food and Environment Protection Act 1985, s 17.

Enforcement

15.1.14 Until recently enforcement of the regulations was entirely the responsibility of officers of the ministers or the Health and Safety Executive. However, in November 1991, the ministers announced that they were to authorise officers of local authorities and port health authorities to enforce the regulations on the advertisement, sale, supply, storage and use of pesticides in retail and wholesale distribution premises, hotels, residential care homes, catering establishments, offices, sports grounds, garden and amenity parks, exhibition centres, race courses and market garden centres. Local authorities are also to be responsible for enforcement where pesticides are used in private dwellings by the occupier.

Authorised officers have powers of entry to land, premises, vehicles, vessels, aircraft, hovercraft or marine structures together with powers to direct that, where they believe an offence has been committed, anything which is in or on the land etc shall be left undisturbed for as long as it appears to be reasonably necessary and that any reasonable remedial and preventive measures be taken.

15.2 Pesticide Residues

The Regulations[1]

15.2.1 It is an offence for any person to leave or cause to be left in any specified food a level of residue exceeding any maximum level applicable to such food. It is claimed that these regulations implement EC directives[2] on the maximum levels of pesticide residues for cereals and for foodstuffs of animal origin. In fact such implementation is not achieved because the directives require that control shall apply from the time the foods are put into circulation. The offence of 'leaving or causing to be left' can, it is submitted, only be committed by the producer who applied the pesticides. Further, the directives require control of residues 'in or on' food whereas the regulations apply only to residues left in any food. These difficulties arise because the enabling powers in s16(1)(k) of the Food and Environment Protection Act 1985 do not permit any wider offence in regulations made under them. Implementation of new and forthcoming EC measures on pesticide residues will require a wider offence applicable at any stage of production or distribution. The powers under s 16 of the Food Safety Act 1990 would appear to be adequate for such implementation.

The regulations also impose maximum residue limits on certain fruits and vegetables but they do not implement a 1976 directive[3] because it was known when the regulations were being drafted that new EC residue limits were being prepared. Instead, residue limits recommended by Codex[4] have been used. In 1990 a new framework directive[5] was adopted and in due course new residue limits for fruits and vegetables will be adopted by the EC Commission.

1 Pesticides (Maximum Residue Levels in Food) Regulations 1988/1378).
2 Council Directives 86/362/EEC and 86/363/EEC (OJ No L 221; 07.08.86).
3 Council Directive 76/895/EEC (OJ No L 340; 09.12.76).
4 Codex Alimentarius Commission.

5 Council Directive 90/642/EEC (OJ No L 350; 14.12.90).

General Offences[1]

15.2.2 Where pesticide residues are present in or on food to an extent to breach the food safety requirement (see Chapter 10) or, perhaps to render the food 'not of the substance demanded' (see Chapter 5) it is open to food authorities to bring proceedings under the Food Safety Act 1990.

1 Food Safety Act 1990, ss 8 and 14.

Seizure or Disposal of Food

15.2.3 If any food has in it a residue level in excess of any maximum residue level relating to the food in question officers of the ministers have the power to seize or dispose of the consignment containing that food or any part of it, or to require that some other person shall dispose of it. Alternatively, they may direct some other person to take such remedial action as appears to the ministers to be necessary.

These powers are of little practical effect because analysis of food for pesticide residues is long, complex and costly. By the time the results of any analysis were to hand the food sampled would be likely to have been sold and cleared.

Sampling and Analysis

15.2.4 The level of residue in a food must be determined as far as practicable in accordance with the procedures laid down in Parts 5 and 6 of the Guide to Codex Recommendations Concerning Pesticide Residues.[1]

1 Food & Agriculture Organisation of the United Nations/World Health Organisation Joint Food Standards Programme; Codex Alimentarius Commission documents CAC/PR5 – 1984 and CAC/PR6 – 1984.

Offences and Defences

15.2.5 Breach of the pesticide residue regulations[1] is an offence against s 16(12) of the Food and Environment Protection Act 1985. That Act contains a general defence of due diligence[2] and creates criminal liability for directors, managers etc of corporate bodies by whose consent, connivance or neglect an offence by the body corporate has been committed.[3]

1 Pesticides (Maximum Residue Levels in Food) Regulations 1988/1378.
2 Food and Environment Protection Act 1985, s 22 (1).
3 Food and Environment Protection Act 1985, s 21 (6).

15.3 Veterinary Residues

Introduction

15.3.1 Concern about the possible effects of growth promoters on human and animal health some 10 years ago resulted in a series of EC measures[1] prohibiting the use of certain substances and restricting the use of others. These measures were quickly implemented in the United Kingdom and development of the law on this subject continued during the mid-1980s. The United Kingdom now has a fairly comprehensive system of legislative control over veterinary residues in meat and meat products, with further improvements in enforcement under consideration. One weakness, however, is that the principal residues Directive[1b] has not yet been extended to include poultry and farmed fish. Until that is done there is a voluntary system of tests for residues in fish but none for poultry.

The regulations briefly summarised in **15.3.2** and **15.3.3** below were made under the Medicines Act 1968. That Act provides for certain defences[2] relating to act or default of another person and warranties – see Chapter 19. The regulations referred to in **15.3.4** below were made under the European Communities Act 1972 and there is a general defence of 'all due diligence' in the regulations.

1 (a) Council Directive 81/602/EEC (OJ No L 222; 07.08.81); (b) Council Directive 86/469/EEC (OJ No L 275; 26.09.86); (c) Council Directive 88/146/EEC (OJ No L 070; 16.03.88); (d) Council Directive 85/358/EEC (OJ No L 191; 23.07.85).
2 Medicines Act 1968, ss 121 and 122.

Sale of Stilbenes and Thyrostatic Substances[1]

15.3.2 The sale or supply of stilbenes or thyrostatic substances for use as an ingredient in the manufacture of medicinal products for administration to animals and the sale or supply of any medicinal product or animal feeding stuff containing them was prohibited from May 1982.

1 Medicines (Stilbenes and Thyrostatic Substances Prohibition) Order 1982/518.

Administration of Stilbenes and Thyrostatic Substances[1]

15.3.3 The administration of stilbenes, thyrostatic substances, or any medicinal product or animal feeding stuff containing them to bovines, swine, sheep, goats, solipeds, poultry and wild ruminants raised on a holding was also prohibited from May 1982.

1 Medicines (Stilbenes and Thyrostatic Substances) Regulations 1982/626.

Hormone Growth Promoters

15.3.4 The administration of hormone growth promoters to the animals mentioned in **15.3.3** above is prohibited from April 1988 except where necessary for therapeutic treatment or for certain other purposes concerning fertility or reproduction.[1]

1 Medicines (Hormone Growth Promoters) (Prohibition of Use) Regulations 1988/705.

Hormonal Substances in Meat

15.3.5 There was continuing concern in 1988 about hormonal residues in animals and consequently in fresh meat. Two new measures were introduced under the Food Act 1984 to provide additional protection for the public. From June 1988, in final implementation of relevant EC directives,[1] regulations[2] prohibited the sale and slaughter of animals to which have been administered hormonal substances other than authorised substances for which, in the case of slaughter for human consumption, the withdrawal period has elapsed since administration.

1 (a) Council Directive 81/602/EEC (OJ No L 222; 07.08.81); (b) Council Directive 86/469/EEC (OJ

No L 275; 26.09.86); (c) Council Directive 88/146/EEC (OJ No L 070; 16.03.88); (d) Council Directive 85/358/EEC (OJ No L 191; 23.07.85).

2 Animals and Fresh Meat (Hormonal Substances) Regulations 1988/ 849.

Examination for Residues

15.3.6 Regulations[1] were made contemporaneously with those mentioned above to provide for the primary analysis of official samples taken from animals and fresh meat, for notification to the owner and the ministers of positive findings of prohibited substances and excess authorised substances and for reference analysis where such findings are challenged by the owner or where the ministers so decide. Provision is also made for animals at farms and elsewhere to be marked and controlled during examinations and for slaughter and disposal in certain circumstances.

The provisions of these regulations are under review and improved controls on all veterinary residues may be expected in 1992.

1 Animals and Fresh Meat (Examination for Residues) Regulations 1988/848.

Chapter 16

Quantity and Price Marking Requirements

16.1 Quantity Marking

16.1.1 The manner in which quantity marking of foods must be done is laid down in regulations made under the Weights and Measures Act 1985.[1] They apply to any container required by the Act[2] to be marked with information as to quantity and the units of measurement to be used in any such information.

1 Weights and Measures Act 1985, Part IV.
2 Weights and Measures (Quantity Marking and Abbreviations of Units) Regulations 1987/1538, as amended.

Marking with Quantity by Measurement

16.1.2 The marking of any container with information as to quantity by measurement (weight, capacity or volume) must comprise the numerical value of the unit of measurement expressed in words or by means of the relevant symbol or abbreviation which may be lawfully used for trade in relation to that unit. Where the numerical value of the unit of measurement is expressed in words, the reference to that unit must be expressed in words and not by means of a symbol or abbreviation. Any metric quantity used in the marking may not be expressed as a vulgar fraction.

If the goods are packed by gross weight where this is permitted, the information as to quantity must include the word 'gross' or the words 'including container' or other words which indicate that the marked weight includes the weight of the container. There can be no abbreviation of the word 'gross'.

It is not necessary to use the word 'net' where goods are packed by net weight but if the word is used it must not be abbreviated.

Legibility and Position of Marking

16.1.3 Any marking of a container with information as to quantity must be easy to understand, clearly legible and indelible. It must be easily visible to an intending purchaser under normal conditions of purchase and must not in any way be hidden, obscured or interrupted by any other written or pictorial matter. If the information is not on the actual container or on a label securely attached to the container, it must be so placed that it cannot be removed without opening the container.

Size of Marking

16.1.4 Except in the case of a catchweight product (ie any product which is not prepacked according to a pre-determined fixed weight pattern, but is packed in varying quantities), where in any marking of any container the quantity by number or the numerical value of a unit of measurement is expressed in figures, all the relevant figures must be at least the height given below. The following table refers to foods which are required or permitted to be sold by weight, capacity, volume or number as the case may be. In the case of number the figure sizes are based on the size of the container which is treated as equal –

(a) in the case of a rectilinear or approximately rectilinear container, to its height, length or width whichever is the greatest;
(b) in the case of a container with curvilinear or approximately curvilinear cross section, its height or maximum diameter, whichever is the greater.

Mass or weight of goods	*Minimum height of figures*
Not exceeding 50 g	2 mm
Exceeding 50 g but not exceeding 200 g	3 mm
Exceeding 200 g but not exceeding 1 kg	4 mm
Exceeding 1 kg	6 mm
Capacity of goods	
Not exceeding 5 cl	2 mm
Exceeding 5 cl but not exceeding 20 cl	3 mm
Exceeding 20 cl but not exceeding 1 L	4 mm
Exceeding 1 L	6 mm
Volume of goods	
Not exceeding 200 cm^3	3 mm
Exceeding 200 cm^3 but not exceeding 1000 cm^3	4 mm
Exceeding 1000 cm^3	6 mm
Number	
Size of container	
Not exceeding 12 cm	2 mm
Exceeding 12 cm but not exceeding 30 cm	3 mm
Exceeding 30 cm but not exceeding 45 cm	4 mm
Exceeding 45 cm	6 mm

Imperial and Metric Units

16.1.5 The marking of containers with information as to quantity must be in metric units but may, in addition, be in imperial units. However, in the case of goods which are still required to be made up in specified metric or imperial units or in imperial units only, the quantity information must be in both metric and imperial units in the manner given in **16.1.6** below. It is expected that all imperial units for quantity marking of foods will be abolished by 1994.

Dual Metric/Imperial Markings

16.1.6 Where a marking of quantity is in both metric and imperial units the following rules apply:

(a) the marking must indicate the quantity in terms of the same measurement in the imperial system and the metric system;

(b) where goods are required to be made up in specified quantities (see **16.2**), the first indication must be the one in which they are made up;
(c) in any other case the marking may indicate the quantity first in either imperial or metric units;
(d) the indications of quantity in imperial and metric units must be of equal size and distinct, but in close proximity to each other, and nothing must be inserted between them;
(e) in the case of cow's milk (any milk in liquid form other than condensed milk or evaporated milk) or cream, any marking of quantity in imperial units must be entirely in terms of the pint, quart or gallon.

Units of Measurement

16.1.7 With the exception of **16.1.6** (e) above, the units of measurement to be used in marking any container with information as to quantity by measurement must be one of the following:

Measurement	*Metric Unit*	*Imperial Unit*
Volume	cubic metre cubic centimetre litre centilitre millilitre	
Capacity	litre centiltre millilitre	gallon quart pint fluid ounce
Mass or Weight	kilogram gram	pound ounce

Symbols and Abbreviations

16.1.8 The following are the symbols and abbreviations which may be used for the quantity marking of food products:

Metric Units:

Unit of Measurement	*Symbol*
litre	l or L
decilitre	dl or dL
centilitre	cl or cL
millilitre	ml or mL
tonne	t
kilogram	kg
hectogram	hg
gram	g
milligram	mg

Imperial Units:

Unit of Measurement	*Abbreviation*
gallon	gal
quart	qt
pint	pt
fluid ounce	fl oz
pound	lb
ounce	oz

In the case of an abbreviation of an imperial unit, the letter 's' may be added, when appropriate, to indicate the plural.

The 'e' Mark

16.1.9 The 'e' mark indicates that the product which bears the mark has been packed in accordance with the average quantity system.[1] Its form is laid down in regulations[2] and when applied to a container it must be at least 3 mm high; be placed in the same field of vision as that of the statement of quantity marked on the package; and be indelible, clearly legible and visible under normal conditions of purchase.

1 Weights and Measures Act 1985, Part V.
2 Weights and Measures (Packaged Goods) Regulations 1986/2049, as amended.

Field of Vision

16.1.10 The indication of quantity on packages of food is required by the Food Labelling Regulations 1984 to appear in the same field of vision (ie to be simultaneously visible) as the name of the food, the indication of minimum durability or 'use by' date, and the alcoholic strength, as appropriate: see **6.4.23**.

16.2 Quantity Requirements for Certain Foods

Introduction

16.2.1 Certain foods are required to be sold by weight or capacity measurement in a certain manner[1,2] and others are required to be made up when prepacked in prescribed quantities.[2]

Cheese, Fish, Fresh Fruits and Vegetables, Meat and Poultry[1] must, generally speaking, be sold by net weight or, subject to certain maximum container weights, by gross weight when not prepacked. When prepacked the container must be marked with an indication of net weight.

Certain cheeses sold by gross weight (subject to maximum container weights) and soft fruit and mushrooms however sold, need not be marked with a statement of weight but the weight must be made known to the buyer before he pays for or takes possession of the food.

There are exemptions for certain cheeses, and certain fruits and vegetables may be sold by count or by the bunch.

Quantities of less than 5 g (25 g in the case of cheese) and more than 5 kg (10 kg in the case of cheese) are exempt.

1 Weights and Measures Act 1963 (Cheese, Fish, Fresh Fruits and Vegetables, Meat and Poultry) Order 1984/1315, as amended.

2 Weights and Measures (Miscellaneous Foods) Order 1988/2040, as amended.

Prescribed Quantities[1]

16.2.2 Certain foods, subject to exemptions, are required to be prepacked only in prescribed quantities. They are:

Foods	*Prescribed Quantities*
Barley kernels, pearl barley, rice (including ground rice and rice flakes), sago, semolina and tapioca	125 g, 250 g, 375 g, 500 g or a multiple of 500 g. Packs of 75 g or less or more than 10 kg exempted
Biscuits, other than wafer biscuits which are not cream filled	100 g, 125 g, 150 g, 200 g, 250 g, 300 g, or a multiple of 100 g. Packs of 85 g or less or more than 5 kg exempted
Bread in form of whole loaf	400 g or a multiple of 400 g*
Cereal breakfast foods in flake form, other than cereal biscuit foods	125 g, 250 g, 375 g, 500 g, 750 g, 1 kg, 1.5 kg or a multiple of 1 kg. Packs of 50 g or less or more than 10 kg exempted
Chocolate products in bar or tablet form of the following reserved descriptions: chocolate, plain chocolate, gianduja nut chocolate, milk chocolate, white chocolate, filled chocolate, cream chocolate; and skimmed milk chocolate	85 g, 100 g, 125 g, 150 g, 200 g, 250 g, 300 g, 400 g or 500 g. Packs of less than 85 g or more than 500 g exempted
Cocoa products of the following reserved descriptions: cocoa, cocoa powder, fat-reduced cocoa, fat-reduced cocoa powder, sweetened cocoa, sweetened fat-reduced cocoa, sweetened fat-reduced cocoa powder, drinking chocolate, fat-reduced drinking chocolate	50 g, 75 g, 125 g, 250 g, 500 g, 750 g and 1 kg. Packs of less than 50 g exempted or more than 1 kg exempted
Coffee, coffee mixtures and coffee bags (contents)	2 oz, 4 oz, 8 oz, 12 oz, 1 lb, 1.5 lb or a multiple of 1 lb; 75 g, 125 g, 250 g, 500 g, 750 g, 1 kg, or a multiple of 500 g. Packs of less than 25 g or more than 5 kg exempted
Coffee extracts and chicory extracts consisting of solid and paste coffee and chicory products	50 g, 100 g, 200 g, 250 g, (for mixtures of coffee extracts and chicory extracts only) 300 g (for coffee extracts only), 500 g, 750 g, 1 kg, 1.5 kg, 2 kg, 2.5 kg, 3 kg or a multiple of 1 kg. Packs of 25 g or less or more than 10 kg exempted

Foods	*Prescribed Quantities*
Dried fruits of any one or more of the following descriptions – apples (including dried apple rings), apricots, currants, dates figs, muscatels, nectarines, peaches, pears (including dried pear rings), prunes, raisins, sultanas and dried fruit salad	125 g, 250 g, 375 g, 500 g, 1 kg, 1.5 kg, 7.5 kg or a multiple of 1 kg. Packs of 75 g or less or more than 10 kg exempted
Dried vegetables of any of the following descriptions – beans, lentils and peas (including split peas)	125 g, 250 g, 375 g, 500 g, 1 kg, 1.5 kg or a multiple of 1 kg. Packs of less than 100 g or more than 10 kg exempted
Edible fats of any of the following descriptions – (a) butter, margarine, any mixture of butter and margarine, and low fat spreads (butter or margarine substitutes); (b) dripping and shredded suet; (c) lard and compound cooking fat substitutes therefor; (d) solidified edible oil (except in gel form)	50 g, 125 g, 250 g, 500 g, or a multiple of 500 g up to and including 4 kg or thereafter a multiple of 1 kg up to and including 10 kg. Packs of 25 g or less or more than 10 kg exempted for (a) only. In other and more cases – less than 5 g or more than 10 kg
Flour, ie flour of bean, maize, pea, rice, rye, soya bean or wheat and flour products of any following descriptions – cake flour, other than cake mixtures and sponge mixtures; cornflour, other than blancmange powders and custard powers; self-raising flour	125 g, 250 g, 500 g, or a multiple of 500 g and in the case of cornflour, in addition 375 g and 750 g. Packs of less than 50 g or more than 10 kg exempted
Honey	2 oz, 4 oz, 8 oz, 12 oz, 1 lb, 1.5 lb or a multiple of 1 lb. Packs of less than 50 g exempted
Jam and marmalade other than diabetic jam or marmalade. Jelly preserves	2 oz, 4 oz, 8 oz, 12 oz, 1 lb, 1.5 lb, or a multiple of 1 lb. Packs of less than 50 g exempted
Milk	(1) $\frac{1}{3}$ pt, $\frac{1}{2}$ pt, or a multiple of $\frac{1}{2}$ pt (2) 200 ml, 250 ml, 500 ml, 750 ml, 1 L, 2 L, of thereafter a multiple of 500 ml. Packs of 50 ml or less exempted
Molasses, syrup and treacle	2 oz, 4 oz, 8 oz, 12 oz, 1 lb, 1.5 lb, or a multiple of 1 lb. Packs of less than 50 g exempted
Oat products – flour of oats; oatflakes and oatmeal	125 g, 250 g, 375 g, 500 g, 750 g, 1 kg, 1.5 kg, or a mutiple of 1 kg. Packs of 50 g or less, or more than 10 kg exempted
Pasta	125 g, 250 g, 375 g, 500 g, or a multiple of 500 g. Packs of 50 g or less exempted

Foods	*Prescribed Quantities*
Potatoes	8 oz, 12 oz, 1 lb, 1.5 lb or a multiple of 1 lb. 500 g, 1 kg, 1.5 kg, 2 kg, 2.5 kg, or a multiple of 2.5 kg up to and including 15 kg, 20 kg or 25 kg. Packs of less than 5 g or more than 25 kg and where the net weight of each potato is not less than 175 g exempted
Salt	125 g, 250 g, 500 g, 750 g, 1 kg, 1.5 kg or a multiple of 1 kg up to and including 10 kg, 12.5 kg, 25 kg or 50 kg. Packs of 100 g or less exempted
Sugar	125 g, 250 g, 500 g, 750 g, 1 kg, 1.5 kg, 2 kg, 2.5 kg, 3 kg, 4 kg or 5 kg. Packs of 100 g or less, or more than 5 kg exempted
Tea in a tea-bag (contents only)	50 g, 125 g, 250 g, 500 g, 750 g, 1 kg, 1.5 kg, 2 kg, 2.5 kg, 3 kg, 4 kg or 5 kg. Packs of 25 g or less, or more than 5 kg exempted
Tea, other than instant or in tea-bag	50 g, 125 g, 250 g, 500 g, 750 g, 1 kg, 1.5 kg, 2 kg, 2.5 kg, 3 kg, 4 kg or 5 kg or when packed in tins or glass or wooden containers in addition 100 g, 200 g, and 300 g. Packs of 25 g or less or more than 5 kg exempted

* Exempted from the prescribed quantity requirements for loaves of bread are loaves weighing 300 g or less; or any sale in pursuance of a contract for the supply of bread for consumption on the premises of the buyer, if the contract provides for each delivery of bread thereunder to be a specified aggregate quantity of not less than 25 kg and for the weighing of the bread on delivery. Not prepacked bread must be made up in the same prescribed quantities but, of course, is exempt from the weight marking requirements.

1 Weights and Measures (Miscellaneous Foods) Order 1988/2040, art 3 and Sch 1 as amended.

Quantity Marking for Prescribed Quantity Foods

16.2.3 All foods made up in the prescribed quantities listed above are required to be marked with an indication of quantity in the prescribed form. Packs of less than 5 g or 5 ml are exempted. For biscuits, sugar, chocolate and cocoa products the exemption is for 50 g or less; for bread where the net weight of each loaf is less than 300 g and the number of items if more than one in the container is marked on the container or is clearly visible and capable of being easily counted through the container. For packs of potatoes where the net weight of each potato is not less than 175 g there is no requirement for a statement of weight provided there is an indication of the number of potatoes in the pack and a statement of the minimum net weight of each potato. The following notes include variations on the quantity marking exemptions given above.

Quantity Marking for Non-Prescribed Quantity Foods and Exempted Foods

16.2.4 Most foods which are not subject to the prescribed quantity requirements are nevertheless required[1] to be marked with an indication of quantity by weight, capacity, volume or number. Additionally, there are special requirements for certain biscuits, shortbreads, caseines and caseinates, cocoa and chocolate products (not in bar or tablet form), coffee and chicory products (other than those above), milk (in small quantities and sold by vending machines) and preserved milk. Non retail sales are controlled, in that there must either be a weighing in the presence of the buyer, or the goods must be packed and marked as for retail sale, or the necessary information as to quantity must be given on trade documents.

1 Weights and Measures (Miscellaneous Foods) Order 1988/2040, as amended.

Multi-packs

16.2.5 There are special requirements for packs containing two or more packs of different goods. They must either be marked with a statement of quantity for the whole, or a statement of the weight of each enclosed pack and their number, or, if the quantity indication on one of the packs and the number contained can be clearly seen through the outer container, that is sufficient.

16.3 The Average Quantity System

The System

16.3.1 Part V of the Weights and Measures Act 1985 implements the EC requirements as to the average quantity system (see **16.1.9** above). Regulations[1] make detailed arrangements for the operation of the system. The Act and the regulations are supplemented by statutory codes of practice.[2]

1 Weights and Measures (Packaged Goods) Regulations 1986/2049, as amended.
2 Code of Practical Guidance for Packers and Importers – Issue No 1 (HMSO: ISBN 0 11 512922 7), as amended; Manual of Practical Guidance for Inspectors (HMSO ISBN 0 11 512501 9) as supplemented (ISBN 0 11 512978 2).

The Duty of Packers and Importers

16.3.2 There are three basic rules imposed on packers and importers by the average quantity system. They are:

1. The actual contents of the packages shall not be less, on average, than the nominal quantity.
2. Not more than 2.5% of the packages may be non-standard, ie have a negative error larger than TNE (tolerable negative error) specified for the nominal quantity.
3. No package may be inadequate, ie have a negative error larger than twice the specified TNE.

The average quantity system, introduced in 1980, has worked well, presenting few problems for industry or trading standards officers. It is accordingly not appropriate to dwell upon it further in this book.

16.4 Deficient Quantity Offences

Short Weight or Measure

16.4.1 It is an offence[1] to fail to sell food products in accordance with the requirements of regulations and orders made under the Weights & Measures Act 1985. It is also an offence[2] to deliver a lesser quantity than that purported to be sold or to make any misrepresentation as to the quantity of goods.

1 Weights and Measures Act 1985, s 25.
2 Weights and Measures Act 1985, s 28 and 29.

Delivering a Lesser Quantity than Corresponds with the Price Charged

16.4.2 It is an offence to deliver to the buyer a lesser quantity of food than corresponds with the price charged.[1] For example, if a food sold from bulk had an indicated unit price (see below) and the price charged for a given quantity of that food was, on checking, found to be too high for the weight delivered, the offence would be committed.

1 Weights and Measures Act 1985, s 28(1)(b).

16.5 Price Indications

Introduction

16.5.1 New requirements[1] as to the indication of prices for foodstuffs were introduced from 1 September 1991 in implementation of EC requirements.[2] The new requirements revoked all existing price marking orders except for that relating to prepacked milk in vending machines[3] and that relating to price indications on catering premises.[4]

Where a person indicates that any foods are or may be for sale by retail he must indicate in writing the selling price (the aggregate sums that the purchaser is required to pay in order to obtain the food). He is also required, subject to exemptions, to indicate in writing the unit price (the price expressed by reference to prescribed units of measurement) where food is sold from bulk, or prepacked in variable quantities and, from 7 June 1995, certain foods prepacked in pre-established quantities. There are special requirements for the indication of unit prices for cheese, meat and milk prepacked in pre-established quantities.

1 Price Marking Order 1991/1382.
2 Council Directive 79/581/EEC (OJ No L 158; 26.06.79), as amended.
3 Price Marking (Pre-packed Milk in Vending Machines) Order 1976/796.
4 Price Marking (Food and Drink on Premises) Order 1979/361.

The Selling Price

16.5.2 The selling price must be indicated either –

(i) in the case of food sold from bulk, by reference either to a unit of measurement or to a single item or to a specified quantity of the food, or
(ii) in the case of other foods, by reference to a single item or to a specified quantity of the food.

Where the selling prices at which different quantities of food may be purchased do not vary in the same proportion to each other as the quantity of the food, the indication of the selling price must include sufficient information to enable the purchaser to calculate what the selling price of any quantity of the food is.

Unit Prices

16.5.3 As indicated above, foods sold from bulk and those prepacked in variable quantities, subject to the exemptions given in **16.5.4** below, must bear an indication of the unit price. From 7 June 1995, foods prepacked in pre-established quantities must be so marked unless they are packed in one of a series of prescribed nominal quantities.

A pre-established quantity means that the food has been packed in such a way that the quantity of goods in the container corresponds to a previously selected quantity. This includes whole birds and poultry portions that are prepacked within defined weight grades, all the birds or portions in each such grade being marked with the same weight.

The ranges of nominal quantities for liquid foods are:

Foods	*Exemptions from Unit Pricing*
Non-sparkling fermented beverages (for example, cider, perry and mead) excluding wine of fresh grapes, grape must and liqueur wines	10 cl, 25 cl, 37.5 cl, 50 cl, 75 cl, 1 L, 1.5 L, 2 L, 5 L. Reusable bottles: 70 cl, $\frac{1}{2}$ pt, $1\frac{1}{3}$ pt, 2 pt
Vermouths and other wines of fresh grapes flavoured with aromatic extracts; liqueur wines	5 cl up to but not including 10 cl, 10 cl, 20 cl, 37.5 cl, 50 cl, 75 cl, 1 L, 1.5 L, 3 L, 5 L. Reusable bottles: 70 cl
Fermented sparkling beverages (for example cider, perry and mead) excluding wine	10 cl, 20 cl, 37.5 cl, 75 cl, 1 L, 1.5 L, 3 L. Reusable bottles: 70 cl, $\frac{1}{2}$ pt, 1 pt, $1\frac{1}{3}$ pt, 2 pt
Beer made from malt, excluding acid beers	25 cl, 33 cl, 50 cl, 75 cl, 1 L, 2 L, 3 L, 4 L, 5 L. Reusable bottles: 70 cl, $\frac{1}{2}$ pt, 1 pt, $1\frac{1}{3}$ pt, 2 pt
Acid beers, gueuze	25 cl, 37.5 cl, 75 cl. Reusable bottles: 70 cl, $\frac{1}{2}$ pt, 1 pt, $1\frac{1}{3}$ pt, 2 pt
Vinegar and substitutes for vinegar	25 cl, 50 cl, 75 cl, 1 L, 2 L, 5 L. Reusable bottles: 70 cl
Olive oils, other edible oils	25 cl, 50 cl, 75 cl, 1 L, 2 L, 3 L, 5 L, 10 L. Reusable bottles: 70 cl
Milk, fresh, not concentrated or sweetened, excluding yoghurt, kephir, curdled milk, whey and other fermented or acidified milk; milk based beverages	20 cl, 25 cl, 50 cl, 75 cl, 1 L, 2 L. Reusable bottles: 70 cl, $\frac{1}{2}$ pt, 1 pt, $1\frac{1}{3}$ pt*, 2 pt
Waters, including spa waters and aerated waters	12.5 cl, 20 cl, 25 cl, 33 cl, 50 cl, 75 cl, 1 L, 1.5 L, 2 L. Reusable bottles: 70 cl

16.5.3 *Quantity and Price Marking Requirements*

Foods	*Exemptions from Unit Pricing*
Lemonade, flavoured spa waters and flavoured aerated waters and other non-alcoholic beverages not containing milk or milk fats excluding fruit and vegetable juices and concentrates	12.5 cl, 20 cl, 25 cl, 33 cl, 50 cl, 75 cl, 1 L, 1.5 L, 2 L. Reusable bottles: 70 cl
Beverages labelled as alcohol-free aperitifs	10 cl. Reusable bottles: 70 cl
Fruit juices (including grape must) or vegetable juices, whether or not containing added sugar, but un-fermented and not containing spirit, fruit nectar	12.5 cl, 20 cl, 25 cl, 33 cl, 50 cl, 75 cl, 1 L, 1.5 L, 2 L. Reusable bottles: 70 cl

* *Note*: Cow's milk in any liquid form other than that of condensed milk (including evaporated milk) or of cream is not permitted to be made up in a quantity of $\frac{1}{3}$ pint.

The range of nominal quantities for solid foods is:

Foods	*Exemptions from Unit Price Marking*
Butter, margarine, emulsified or non-animal and vegetable fats (low fat spreads)	125 g, 250 g, 500 g, 1 kg, 1.5 kg, 2 kg, 2.5 kg, 5 kg
Fresh cheeses except 'petits suisses' and other cheeses put up in the same way	62.5 g, 125 g, 250 g, 500 g, 1 kg, 2 kg, 5 kg
Table and cooking salt	125 g, 250 g, 500 g, 750 g, 1 kg, 1.5 kg, 5 kg
Impalpable sugars, red or brown sugars, candy sugars	125 g, 250 g, 500 g, 750 g, 1 kg, 1.5 kg, 2 kg, 2.5 kg, 3 kg, 4 kg, 5 kg
Cereal flours, groats, flakes and oatmeal and meal, oatflakes, (excluding prepared foods obtained by the swelling or roasting of cereals or cereal products (puffed rice, cornflakes and similar products))	125 g, 250 g, 500 g, 1 kg, 1.5 kg, 2 kg, 2.5 kg (not applicable to oatmeal and oatflakes), 5 kg, 10 kg
Prepared foods obtained by the swelling or roasting of cereals or cereal products (puffed rice, cornflakes and similar products)	250 g, 375 g, 500 g, 750 g, 1 kg, 1.5 kg, 2 kg
Pasta products	125 g, 250 g, 500 g, 1 kg, 1.5 kg, 2 kg, 3 kg, 4 kg, 5 kg, 10 kg
Rice	125 g, 250 g, 500 g, 1 kg, 2 kg, 2.5 kg, 5 kg
Dried vegetables (excluding potatoes and dehydrated vegetables); dried fruits	125 g, 250 g, 500 g, 1 kg, 1.5 kg, 2 kg, 5 kg, 7.5 kg, 10 kg
Ground or unground roasted coffee, chicory and coffee substitutes	125 g, 250 g, 500 g, 1 kg, 2 kg, 3 kg, 4 kg, 5 kg, 10 kg
Frozen fruit and vegetables and pre-cooked potatoes for chips	150 g, 300 g, 450 g, 600 g, 750 g, 1 kg, 1.5 kg, 2 kg, 2.5 kg

Foods	*Exemptions from Unit Price Marking*
Frozen fish fillets and portions, breaded or not breaded	100 g, 200 g, 300 g, 400 g, 500 g, 600 g, 800 g, 1 kg, 2 kg
Frozen fish fingers	150 g, 300 g, 450 g, 600 g, 900 g, 1.2 kg, 1.5 kg, 1.8 kg
Ice cream in quantities of more than 250 ml (excluding any ice-cream whose volume is not determined by the shape of the container)	300 ml, 500 ml, 750 ml, 1 L, 1.5 L, 2 L, 2.5 L, 3 L, 4 L, 5 L

Other foods are required to be unit priced but have exemptions based on container capacities. They are:

Foods	*Exemptions from Unit Pricing*
Preserves and semi-preserves in tins and glass containers:	
(a) Vegetable products (fruit, vegetables, tomatoes, potatoes, with the exception of asparagus, soups, fruit or vegetable juices and fruit nectars) for human consumption (excluding a description mentioned in any of the following items)	106 ml, 156 ml, 212 ml, 228 ml, 314 ml, 370 ml, 425 ml, 446 ml, 580 ml, 720 ml, 850 ml, 1.062 L, 1.7 Lm 2.65 L, 3.1 L, 4.25 L, 10.2 L
(b) Truffles	26 ml, 53 ml, 71 ml, 106 ml, 212 ml, 425 ml, 720 ml, 850 ml
(c) Tomatoes: – concentrates	71 ml, 142 ml, 212 ml, 370 ml, 425 ml, 720 ml, 850 ml, 3.1 L, 4.25 L
– peeled or unpeeled	236 ml, 370 ml, 425 ml, 720 ml, 850 ml, 2.65 L, 3.1 L
(d) Fruit cocktails, fruits in syrup	106 ml, 156 ml, 212 ml, 228 ml, 236 ml, 314 ml, 370 ml, 425 ml, 446 ml, 580 ml, 720 ml, 850 ml, 1.062 L, 1.7 L, 2.65 L, 3.1 L, 4.25 L, 10.2 L

It will be noted that the exempted quantities listed above do not correspond in all cases to the prescribed quantities listed at **16.2**. It may be expected that prior to the commencement date for unit pricing, ie 7 June 1995, new EC ranges of prescribed quantities will be introduced to eliminate this anomaly.

Intoxicating liquors are required to be unit priced, where, under specified circumstances they are not made up in one of the quantities prescribed by the Weights and Measures (Intoxicating Liquor) Order 1988.[1]

Cheese, meat and milk. These foods were previously required to be unit priced by orders in force before this Order came into force. Their provisions have been re-enacted and continue the requirement that specified foods within these categories, subject to exemptions, shall bear an indication of unit price.

1 Weights and Measures (Intoxicating Liquor) Order 1988/2039, as amended.

Exemptions from Unit Pricing

16.5.4 The Order[1] provides for exemptions from the requirement to indicate unit prices for:

(a) foods not required by Parts IV or V of the Weights and Measures Act 1985 to be sold by quantity or marked with an indication of quantity;
(b) foods sold by the bunch or by number;
(c) an assortment of different items sold in a single pack;
(d) foods sold from vending machines;
(e) prepared dishes of food or foods sold in a single pack from which a mixture is to be prepared;
(f) fancy foods made in the form of a figurine, animal, cigarette or egg or in any other fancy form;
(g) multi-packs of exempted individual packs;
(h) a portion of meat or poultry cut at the request of, and in the presence of, an intending purchaser;
(i) certain foods packed in aerosols;
(j) intoxicating liquors and other liquids packed in quantities above or below the limits set in the Weights and Measures (Intoxicating Liquor) Order 1988;[2]
(k) food sold in quantities of less than 50 g or 50 ml or more than 10 kg or 10 L;
(l) prepacks in the range of pre-established quantities listed at **16.5.3** above, which are sold on premises where the number of employees who are each normally required to work there for 16 or more hours a week does not exceed two, and which the seller (or his servant or agent) ordinarily himself selects and hands to the purchaser or delivers as the purchaser may direct, and certain aerosol products.

1 Price Marking Order 1991/1382, art 4.
2 Weights and Measures (Intoxicating Liquor) Order 1988/2039, as amended.

Units of Measurement for Unit Pricing

16.5.5 Where a unit price is required it must be expressed in units of measurement laid down by the Order. For fruits and vegetables, meat, fish and cheese (all subject to exemptions) the units are the pound or the kilogram. Thus a unit price for a cut of fresh beef would be expressed as, for example, '£2.50 per pound' or '£5.62 per kilogram'.

For all other foods which are or will be required to be unit priced the units of measurement are:

Kilogram	pound
100 gram (g)	quarter pound ($\frac{1}{4}$ lb)
litre	gallon
100 millilitre (ml)	pint

Imperial Equivalent Units

16.5.6 A surprise requirement of the Order[1] was that where a unit price is given in metric units of measurement for foods sold from bulk or in variable quantities, or for cheese, meat and milk, it must be accompanied by an indication in imperial units. Those imperial units must be, in the case of cheese, meat, milk and fish, the pound and in any other case the pound or quarter pound as an equivalent to the kilogram or 100 g, and the pint or gallon as an equivalent to 100 ml or the litre.

1 Price Marking Order 1991/1382, art 3.

Manner of Price Marking

16.5.7 An indication of price or the display of a statement required by the Order[1] must be unambiguous and easily identifiable by the prospective purchaser as referring to the goods in question and clearly legible.

The indication of the selling price when food is exposed for sale must be given in any one or more of the following ways:

(a) on the food, or if the food is prepacked, on the container;
(b) on a ticket or notice displayed on or in close proximity to the food to which it refers;
(c) grouped together with other prices on a list in close proximity to the food to which it refers.

The indication of unit price must be given in one or both of the following ways:

(a) on or adjacent to the food or, if the food is prepacked, on or adjacent to the container;
(b) together with other indications of the unit prices of other foods in the form of a list or otherwise.

If the unit price has been calculated by reference to gross weight, where sale by gross weight is permitted, that fact must be clearly indicated.

1 Price Marking Order 1991/1382, art 8.

Perishable Foods Reduced in Price to Clear

16.5.8 A problem under revoked orders requiring unit pricing was that a reduction in the selling price would make the indicated unit price inaccurate and thus unlawful. That has been rectified in that perishable foods sold at reduced prices because of the danger of their deterioration are now exempted from the need to bear a unit price.[1]

1 Price Marking Order 1991/1382, art 4(3)(i).

16.6 Misleading Indications of Price

16.6.1 It is an offence to give any indication of price which is misleading.[1] Guidance as to misleading price indications has been provided by a statutory Code of Practice[2] but it is generally agreed that control of misleading prices by way of the Code of Practice has been less than satisfactory. A Departmental Working Party has recommended that the Code, in whole or in part, should be replaced by regulations. Ministers are believed to be sympathetic to the recommendations.

1 Consumer Protection Act 1987, s 20.
2 Code of Practice for Traders on Price Indications (1988) Department of Trade and Industry.

Chapter 17

Enforcement, Sampling, Analysis and Examination

17.1 EC Requirements

Official Control of Foodstuffs

17.1.1 The Council Directive on the Official Control of Foodstuffs,[1] (in this chapter referred to as 'the Directive') is the principal instrument from which the enforcement of food law in the EC is derived. In the United Kingdom the essential requirements of the Directive have been implemented by the Food Safety Act 1990[2] and regulations made thereunder.[3] In certain respects, which are identified below, it is at least arguable that the United Kingdom has not yet implemented the Directive in its entirety. However, there can be no doubt that the arrangements for the enforcement of Community food law in the United Kingdom are more effective and more comprehensive than in any other member state.

The Directive does not apply to metrological controls which, in the United Kingdom, are enforced under the Weights and Measures Act 1985:[4] see **17.5** below.

In the following paragraphs the requirements of the Directive are summarised and a reference as to the United Kingdom's implementation of them is given.

1 Council Directive 89/397/EEC (OJ No L 186; 30.06.89).
2 Food Safety Act 1990, Part III and ss 40, 41, and 42.
3 (a) Food Safety (Enforcement Authority) (England and Wales) Order 1990/2462; (b) Food Safety (Sampling and Qualifications) Regulations 1990/2463; (c) Food Safety (Exports) Regulations 1991/1476.
4 Weights and Measures Act 1985, ss 79–83.

Objectives of the Directive

17.1.2 The Directive requires member states to appoint competent authorities for the purposes of food control. In the United Kingdom those competent authorities are the ministers and local food authorities (see Chapter 2). The function of those authorities is to secure the compliance of foodstuffs; of food additives, vitamins, mineral salts, trace elements and other additives intended to be sold as such; and of materials and articles intended to come into contact with foodstuffs with provisions aimed at preventing risks to public health, guaranteeing fair commercial transactions or protecting consumer interests, including provisions on consumer information. For a list of EC law applicable to these objectives, see Appendix B.

The Directive applies the above objectives to consignments of food to another member state and also requires that products should not be excluded from official control on the grounds that they are intended for export outside the Community.[1]

1 Food Safety (Exports) Regulations 1991/1476.

General Inspection Requirements

17.1.3 The Directive requires that inspections shall be carried out regularly and where non-compliance is suspected. The means of inspection must be proportionate to the end to be observed and must cover all stages of production, manufacture, import into the Community, processing, storage, transport, distribution and trade.[1] It should be carried out at the stage most appropriate to achieve the objective. As a general rule, inspections should be carried out without prior warning.[2]

1 Food Safety Act 1990, s 32.
2 Code of Practice No 3: Inspection Procedures General (HMSO ISBN 0 11 321355 7).

Operations of Control

17.1.4 Control must include one or more operations of inspection; sampling and analysis; inspection of staff hygiene; examination of written and documentary material; and examination of any verification systems set up by the undertaking and of the results obtained.[1,2,3]

1 Food Safety Act 1990, s 32.
2 Code of Practice No 3: Inspection Procedures General (HMSO ISBN 0 11 321355 7); Code of Practice No 8: Food Standards Inspections (HMSO ISBN 0 11 321466 9); Code of Practice No 9: Food Hygiene Inspections (HMSO ISBN 0 11 321467 7); Code of Practice No 2: Legal Matters (HMSO ISBN 0 11 321353 0).
3 Food Safety Act 1990, s 21.

Content of Inspection

17.1.5 The following must be subject to inspection:

(a) the state and use which is made at the different stages of production and distribution of the site, premises, offices, plant surroundings, means of transport, machinery and equipment;
(b) raw materials, ingredients, technological aids and other products used for the preparation and production of foodstuffs;
(c) semi-finished products;
(d) finished products;
(e) materials and articles intended to come into contact with foodstuffs;
(f) cleaning and maintenance products and processes and pesticides;
(g) processes used for the manufacture or processing of foodstuffs;
(h) labelling and presentation of foodstuffs;
(i) preserving methods.

These operations may be supplemented by interviews with the head of undertakings or persons working therein; the reading of values recorded by measuring instruments installed by the undertaking; and inspections carried out by the competent authority, with its own instruments, of measurements taken with the instruments installed by the undertaking.[1,2]

1 Food Safety Act 1990, Part III and ss 40, 41, and 42.
2 Code of Practice No 3: Inspection Procedures General (HMSO ISBN 0 11 321355 7).

Sampling and Analysis

17.1.6 During inspection samples of any of the items in **17.1.5** (b)–(f) may be taken for analysis and analyses must be carried out at official laboratories.[1,2]

1 Food Safety Act 1990, ss 27, 28, 29, 30 and 31.
2 Food Safety (Sampling and Qualifications) Regulations 1990/2463.

Inspection of Persons

17.1.7 Persons who, in the exercise of their activity, come into contact, whether directly or indirectly, with the materials and products referred to in **17.1.5** (b)–(f) are subject to hygiene inspection. The purpose of this inspection is to check that health standards concerning personal cleanliness and clothing are respected. Such inspection is without prejudice to medical examinations.[1]

1 Food Hygiene (Docks, Carriers etc) Regulations 1960/1602, Part III; Food Hygiene (Markets, Stalls and Delivery Vehicles) Regulations 1966/791, Part III; Food Hygiene (General) Regulations 1970/1172, Part III; – see also Chapter 12.

Examination of Documents

17.1.8 Inspectors may take note of written and documentary material held by the natural and legal persons[1] at the various stages in manufacture and distribution. They may also make copies or take extracts of written and documentary material with them for examination.[2,3]

1 A 'natural person' is an individual, whether a proprietor, director, manager or employee. A 'legal' person is a body corporate.
2 Code of Practice No 3: Inspection Procedures General (HMSO ISBN 0 11 321355 7); Code of Practice No 8: Food Standards Inspections (HMSO ISBN 0 11 321466 9); Code of Practice No 9: Food Hygiene Inspections (HMSO ISBN 0 11 321467 7); Code of Practice No 2: Legal Matters (HMSO ISBN 0 11 321353 0).
3 Food Safety Act 1990, s 32 (5) and (6).

Obstruction

17.1.9 Member states must ensure that inspectors have the right to carry out the operations provided for in the Directive and that natural and legal persons[1] concerned are obliged to undergo any inspection carried out in accordance with the Directive and to assist inspectors in the accomplishment of their tasks.[2]

1 A 'natural person' is an individual, whether a proprietor, director, manager or employee. A 'legal' person is a body corporate.
2 Food Safety Act 1990, s 33.

Legal Action and Appeals

17.1.10 Where inspectors discover or suspect an irregularity they must take the requisite measures.[1] Member states must make provision for appeals against measures taken by the competent authority[2] and prescribe that inspectors are bound by professional secrecy.[3]

1 Food Safety Act 1990, ss 9, 10, 12, 35 and 36.
2 Food Safety Act 1990, ss 37, 38, and 39; Magistrates' Courts Act 1980, s 111 (2); Magistrates' Courts Rules 1981/552, rr 76 and 81.
3 Food Safety Act 1990, s 32 (7).

Inspection Programmes

17.1.11 The Directive requires member states to draw up forward programmes of inspections laying down the nature and frequency of inspection to be carried out regularly.[1,2] By 1 May in each year the member states must submit reports on implementation during the previous year specifying:

- the criteria applied in drawing up these programmes,
- the number and type of inspections carried out,
- the number and type of infringements established.

By 16 October in each year (beginning in 1991) the Commission is required to transmit to the member states, having consulted them, a recommendation concerning a coordinated programme of inspections. This work is carried out in consultation with the Standing Committee for Foodtuffs on recommendations submitted by a team of experienced enforcement officers, one being drawn from each member state.

1 (a) Code of Practice No 3: Inspection Procedures General (HMSO ISBN 0 11 321355 7); (b) Code of Practice No 8: Food Standards Inspections (HMSO ISBN 0 11 321466 9); (c) Code of Practice No 9: Food Hygiene Inspections (HMSO ISBN 0 11 321467 7); (d) Code of Practice No 2: Legal Matters (HMSO ISBN 0 11 321353 0).
2 Food Safety Act 1990, s 41.

Competent Authorities and Laboratories

17.1.12 Member states are required to send to the Commission the names of the competent authorities and the extent of their territorial responsibility and functions together with the official laboratories authorised by the competent authorities, which are responsible for carrying out analyses in connection with control. These lists are published in the 'C' series of the Official Journal of the EC.

Practical Effect of the Directive

17.1.13 Enforcement officers derive their powers from the Food Safety Act 1990 but they and persons in the food industry should bear in mind the requirements of the Directive. Although not directly binding, the Directive is an expression of Community law and could, in the ultimate, be enforced by the European Court of Justice. Further, the ministers have the power[1] to extend powers of enforcement in accordance with Community requirements.

1 Food Safety Act 1990, ss 16 and 17.

17.2 Powers of Entry

Power to Enter Premises

17.2.1 The Food Safety Act 1990 provides[1] that authorised officers of an enforcement authority have the right, on producing, if required, some duly authenticated document[2] showing their authority, and at all reasonable hours –

(a) to enter any premises within the authority's area for the purpose of ascertaining whether there is or has been on the premises any contravention of the provisions of the Act, or regulations or orders made under it; and
(b) to enter any business premises, whether within or outside the authority's area, for the purpose of ascertaining whether there is on the premises any evidence of any contravention within that area of any of such provisions; and
(c) in the case of an authorised officer of a food authority, to enter any premises for the purpose of the performance by the authority of their functions under the Act.

It is clear that an authorised officer may enter any premises, whether domestic or business premises within his own authority's area but may only enter business premises outside that area. The power to enter business premises in the area of another food authority has been provided to facilitate visits to, for example, a manufacturer as part of an investigation into an offence concerning one of his products found in the area of the visiting officer. It is recommended, however, that such visits should be coordinated with the home authority and should not include any advice or recommended changes to procedures etc which more properly fall within the remit of the home authority.[3]

Entry to premises used only as a private dwelling house may not be demanded as of right unless 24 hours' notice of the intended entry has been given to the occupier. However, a dwelling house used as part of a food business would be regarded as business premises.

As to 'reasonable hours' see **10.2.2.**

1 Food Safety Act 1990, s 32.
2 Food Safety Act 1990, s 49.
3 Code of Practice No 3: Inspection Procedures – General (HMSO ISBN 0 11 321355 7).

Resisted Entry

17.2.2 If admission to premises has been refused, or refusal is apprehended, an authorised officer may, on sworn information in writing and having given notice to the occupier of the premises, apply to a justice of the peace for a warrant authorising the officer to enter the premises, if need be by force. If an application for admission, or the giving of notice, would defeat the object of entry, or in a case of urgency, or where the premises are unoccupied or the occupier is temporarily absent, the justice may nevetheless issue the warrant. Such a warrant continues in force for one month.

Assistance and Coordination

17.2.3 Authorised officers entering premises may take other persons with them as they consider necessary for the carrying out of their duties. Codes of practice[1] recommend coordination of inspection visits to include, where necessary, multi-disciplinary teams of officers.

1 Code of Practice No 3: Inspection Procedures – General (HMSO ISBN 0 11 321355 7); Code of Practice No 8: Food Standards Inspections (HMSO ISBN 0 11 321466 9); Code of Practice No 9: Food Hygiene Inspections (HMSO ISBN 0 11 321467 7); Code of Practice No 2: Legal Matters (HMSO ISBN 0 11 321353 0).

Computers, Records and Other Documents

17.2.4 On entering premises in accordance with the Act an authorised officer may inspect any records, in whatever form they are kept, relating to a food business and where such records are kept by means of a computer he must have access to them and may require a print-out of them. Further, where any records, documents or computer print-out are required as evidence the officer may seize and detain such records. Alternatively, the officer may take copies of any documents etc, giving notice to the owner that the originals may be required as evidence.

Assistance or Information

17.2.5 An authorised officer may seek any other reasonable assistance or information in connection with the food business under inspection.[1] The Food Safety Act 1990 does not specifically provide the powers required to exercise the functions (other than sampling) required by the Directive discussed under **17.1.5** above. The Act does provide that entry is for the purpose of ascertaining whether there has been any contraventions of the Act, regulations or orders, and the reference to reasonable assistance may in fact cover the point. It is, however, surprising that specific powers to cover all inspection requirements of the Directive were not provided. This is, perhaps, a matter for the ministers to resolve by regulations.

1 Food Safety Act 1990, s 33 (1) (b).

Inspection of Due Diligence Systems

17.2.6 There is no power for authorised officers to inspect due diligence control systems being maintained by a food business, to provide for a defence under the Act.[1] The powers are linked to the need to ascertain whether there has been any contravention of the Act and there is no requirement under the Act for any food business to have a due diligence system. However, if the enforcement officer has reason to believe that an offence has been committed he may, and indeed should, satisfy himself whether or not the control systems operated by the business would in fact amount to a defence. The powers as to records and documents discussed above would assist him in so doing.

1 Food Safety Act 1990, s 21.

Obstruction

17.2.7 It is an offence for any person to intentionally obstruct any person acting in the execution of the Act or, without reasonable cause, to fail to give to any such person any assistance or information which he may reasonably require for the performance of his functions under the Act. Further, it is an offence for any person responding to a request for information to furnish information which he knows to be false or misleading in a material particular; or to recklessly furnish information which is false or misleading in a material particular.[1]

Obstruction need not involve physical violence.[2] Anything which might make the task of the authorised officer more difficult may amount to obstruction.[3]

1 Food Safety Act 1990, s 33 (1), (2).
2 *Borrow v Howland* (1896) 74 LT 787; *Betts v Stevens* [1910] 1 KB 1.
3 *Hinchcliffe v Sheldon* [1955] 3 All ER 406, [1955] 1 WLR 1207; *Rice v Connolly* [1966] 2 QB 414, [1966] 2 All ER 649.

Self-incrimination

17.2.8 The offence of failing, without reasonable cause, to give an authorised officer reasonable assistance or information does not require any person to answer any questions or give any information if to do so might incriminate him.[1] This is an expression of the principle that 'no person should be convicted out of his own mouth'. Although there is no judicial authority for saying so, it is submitted that this safeguard does not justify a refusal to allow the inspection of, for example, recipes. Where there is a statutory power conferred on an authorised officer he must not be prevented from exercising that power. On the other hand, it would not be an offence, by virtue of this provision, for a person to refuse to answer questions about his or her alleged failure to comply with company instructions. For a further discussion of self-incrimination see Chapter 18.

1 Food Safety Act 1990, s 33 (3).

Reports on Inspections

17.2.9 Although there is no statutory requirement for it, the code of practice[1] recommends that authorised officers should report back to the owner of the food business in writing. The report should give the date, time and place of the inspection, what was examined, whether any samples were taken, and the officer's conclusions. This is a most helpful development. Such report should be retained by the food business for incorporation into the due diligence system (see Chapter 19).

1 Code of Practice No 3: Inspection Procedures – General (HMSO ISBN 0 11 321355 7).

17.3 Sampling and Analysis

Procurement of Samples

17.3.1 An authorised officer of an enforcement authority may –

(a) purchase a sample of any food, or any substance capable of being used in the preparation of food,
(b) take a sample of any food which –
  (i) appears to him to be intended for sale, or to have been sold, for human consumption; or
  (ii) is found by him on or in any premises which he is authorised to enter,
(c) take a sample from any food source, or a sample of any contact material, which is found by him on or in any such premises,
(d) take a sample of any article or substance which is found by him on or in any such premises and which he has reason to believe may be required as evidence in proceedings under any of the provisions of this Act or of regulations or orders made under it.[1]

This very wide power enables an authorised officer of a food authority to obtain a sample of any food etc in almost any conceivable circumstance in relation to the manufacture, processing, importation or distribution of food etc. It is for the officer to decide whether to purchase the same or take it. However, he must bear in mind that 'take' does not include 'purchase',[2] and follow the correct course in the light of what he is trying to achieve. For example, he cannot purchase a sample of a contact material.

The term 'sample' is not defined but an item of food handed to an authorised officer is not a sample[3] because it has not been procured in accordance with the above provisions. This does not, of course, prevent an item of food which is the subject of a complaint from being submitted for analysis or examination. For the handling of consumer complaints see Code of Practice No 2: Legal Matters.[4] A purchaser of an item of food may also submit it for analysis or examination – see **17.3.2** below.

1 Food Safety Act 1990, s 29.
2 *Marston v Wrington Vale Dairies Ltd* (1963) 61 LGR 202.
3 *Arun District Council v Argyle Stores Ltd* (1986) 150 JP 552; *Love v Strickland and Holt Ltd* (1981) unreported; and *Leach v United Dairies Ltd* [1949] 1 All ER 1023.
4 Code of Practice No 2: Legal Matters (HMSO ISBN 0 11 321353 0).

Analysis etc of Samples

17.3.2 An authorised officer who has procured a sample may, but is not obliged to, submit it for analysis by a public analyst or, if he considers that it should be examined, submit it to a food examiner,[1] (see **2.6** and **2.7**). The public analyst to whom the sample is submitted can be either one appointed by the food authority for the area where the sample was procured or one appointed for the area of the food authority concerned. The first alternative allows for the submission of a sample procured by an authorised officer outside his own area to the public analyst for the area where the sample was procured.

A person other than an authorised officer who has purchased any food or substance capable of being used in the preparation of food may submit a sample of it to the public analyst for the area where the purchase was made or to a food examiner.

A food examiner to whom a sample is submitted may be any food examiner who holds the necessary qualifications (see **2.7**). In practice most public analysts are also food examiners, or they employ someone who is.

There is nothing to prevent a food authority from bringing proceedings in respect of a sample which has not been submitted to a public analyst. It may be, for example, that identification of a species or cut of a food is all that is required for a successful prosecution and expert testimony can be obtained from a person other than a public analyst, or that there was an apparently sterile foreign body in the food which does not require analysis. In practice, however, it is only in exceptional circumstances that the advice of the public analyst, who is the food authority's scientific adviser, should not be sought.

1 Food Safety Act 1990, s 30.

Food Analysts and Food Examiners

17.3.3 As discussed in Chapter 2, a food analyst is a public analyst or any other person who holds the requisite qualifications to carry out analyses for the purposes of the Food Safety Act 1990, namely a Mastership in Chemical Analysis of the

Royal Society of Chemistry.[1] A food examiner is a person who holds the requisite qualifications (ie one of a number of academic qualifications and has prescribed experience).[1,2] Food analysts and examiners have a particularly important role in the food industry in contributing to due diligence defence systems (see Chapter 19) but for the purposes of this chapter reference to a food analyst means a public analyst and reference to a food examiner means one who is carrying out food examinations for a food authority.

1 Food Safety (Sampling and Qualifications) Regulations 1990/2463.
2 Food Safety Act 1990, s 30 (9).

Submission of Samples to Food Analysts and Food Examiners

17.3.4 A food analyst or food examiner is required to analyse or examine a sample submitted to him as soon as practicable. It is particularly important that this be done and that samples be submitted by the authorised officer quickly in relation to short life foods. After analysis or examination the food analyst or food examiner must give a certificate in the prescribed form[1] to the person by whom the sample was submitted, specifying the result of the analysis or examination. The analysis or examination may be carried out by a person under the supervision of the food analyst or food exmainer.[2]

1 Food Safety (Sampling and Qualifications) Regulations 1990/2463.
2 Food Safety Act 1990, s 30 (5), (6) and (7).

Evidential Status of Certificates of Analysis or Examination

17.3.5 In any proceedings under the Food Safety Act 1990, the production by one of the parties of (i) a document purporting to be a certificate given by a food analyst or examiner, or (ii) of a document supplied to him by the other party as being a copy of such a certificate, is sufficient evidence of the facts stated unless in the case of (i) above the other party requires that the food analyst or examiner shall be called as a witness.[1] This applies to any certificate given by a food analyst or examiner to a food business which is produced by that business in defence of a charge under the Act.

The evidence of a food analyst or food examiner may be rebutted. Where there is doubt in any proceedings about the correctness of the analysis or examination or as to any opinions offered on a certificate, notice should always be given to the other party that the food analyst or food examiner is required to give evidence.

Methods of analysis are increasingly laid down in EC law (see Appendix B). Where there are no prescribed methods of analysis, food analysts and food examiners carry out their work to the highest scientific standards and by the most reliable methods available at the time. Recommendations are issued by the relevant committees of the Royal Society of Chemistry on such methods and standards.

1 Food Safety Act 1990, s 30 (8).

Reference to the Government Chemist

17.3.6 Where an authorised officer and an owner of a retained part of a sample agree, or where a court so orders, the authorised officer must submit it to the Government Chemist or to such other food analyst as the Government Chemist may direct. This procedure is used, for example, where there is a difference between the result of the analysis carried out by the public analyst and that carried out by another food analyst on the part of the sample given to the owner.

17.4 Treatment of Samples

Principles of Fair Sampling

17.4.1 Certain basic principles of fair sampling have evolved as food law has been developed during the 20th century. They may be summarised as:

(a) *The sample should be representative of the food.* The treatment of a sample should be such that no false impression is created of the food by its selection for or method of sampling;
(b) *The sample should be divided into three parts.* The sample should be divided into three equal and representative parts in the presence of the owner if he so wishes. He should be allowed to choose one part of the finished sample, another part should be submitted for analysis or examination, as appropriate, and the third part should be retained by the authorised officer;
(c) *Each part of the sample should be sealed and marked.* The sample should be adequately sealed to prevent tampering by either party to proceedings after the division is complete. The sample should be marked with a number, a description and identification of the food authority;
(d) *The retained part of the sample must be held in safe keeping by the authorised officer and produced in court.* If a prosecution results from the sample it is important that the retained part be produced so that the court is aware of its existence and that it may be submitted to the government chemist;
(e) *The owner of the food should be informed that the sample is to be submitted for analysis.* This is to alert him to the possibility of proceedings so that he may retain relevant documents, notify other parties such as the manufacturer, take parallel samples from stock for future reference and, perhaps, have the part of the sample left with him (the reference part) analysed.

These principles have been modified over the years to permit a sample to consist of unopened containers in cases where the opening of the containers may inhibit a proper analysis. A further problem has been the sampling of products of a non-homogenous nature such as sausage rolls and meat pies. It was near impossible to obtain a fair and representative sample of such a product no matter what lawful sampling procedure was used.[1]

1 *Skeate v Moore* [1971] 3 All ER 1306, [1972] 1 WLR 110.

Average Sampling

17.4.2 A single item of food may be the subject of an offence against the Food Safety Act 1990[1] no matter how unrepresentative it may be of other products of the same kind on sale at the same time. This has been a long-standing problem in food law administration, hence the recommendation that the owner of food from which a sample has been procured should take parallel samples from the same consignment or batch for independent analysis. If such analysis indicated that the official sample was indeed unrepresentative, the court could be acquainted of the facts. This would not be a defence to a charge but it would be a powerful submission in mitigation.

Conscious of this problem the Ministry of Agriculture, Fisheries and Food and the EC Commission have been considering the practicality of a system of average sampling and analysis. Such a system would be very expensive to operate and it is unlikely that such a system will be introduced in the forseeable future. To a certain extent the problem has been overcome by in-factory enforcement because authorised officers now have access to recipes and can observe the introduction of ingredients to the mixing bowl.

1 Food Safety Act 1990, ss 7, 8, 14 and 15.

New Sampling Requirements for Analysis

17.4.3 It had been hoped that new regulations[1] and codes of practice[2] made under the Food Safety Act 1990 would resolve most of these problems. Unfortunately they have created more problems than they have solved. The regulations are badly drafted and the code of practice is unhelpful. Both fall far below the high standards achieved in all other food legislation made during 1990–91.

The prescribed methods of sampling now in force are:

(a) An authorised officer who has procured a sample and who considers that it should be analysed must forthwith divide that sample into three parts.

(b) If the sample consists of sealed containers and opening them would, in the opinion of the authorised officer, impede a proper analysis, the authorised officer must divide the sample into parts by putting the containers into three lots, and each lot must be treated as being a part.

(c) The authorised officer must –
   (i) if necessary, place each part in a suitable container and seal each container;
   (ii) mark each part or container;
   (iii) as soon as it is reasonably practicable to do so, give one part to the owner and give him notice that the sample will be analysed;
   (iv) retain one part for future submission to the government chemist.

   Circumstances where foods may be best left in unopened containers include –
   (i) foods containing evanescent ingredients eg vitamins which disappear quickly after opening;
   (ii) foods containing volatile substances such as alcohol;
   (iii) products where it is not possible to ensure that each ingredient is divided equally between the three parts such as products carrying a declaration of added water, meat products with lumps of meat and products containing pieces of fruit or nuts;
   (iv) products where it is difficult to extract the whole of the contents from the container;
   (v) foods packed in aerosols;
   (vi) 'aerated' foods;
   (vii) products in hermetically sealed containers.

(d) If the authorised officer is of the opinion that division of the sample into parts is either not reasonably practicable or is likely to impede a proper analysis, he must as soon as is reasonably practicable to do so give to the owner notice that it will be analysed and then submit it for analysis. This procedure is to be used in cases where there is insufficient product available; where there are problems of storage such as with frozen products; and where foods are not homogeneous as discussed at **17.4.1** above.

By an apparent drafting oversight a sample taken under this procedure need not be sealed or marked. This could have serious consequences for both the prosecution and defence of proceedings and it is recommended that the sample should be treated in the same manner as a sample procured for examination – see **17.4.4** below.

1 Food Safety (Sampling and Qualifications) Regulations 1990/2463.
2 Code of Practice No 7: Sampling and Analysis or Examination (HMSO ISBN 0 11 321351 4).

New Requirements as to Sampling for Examination

17.4.4 An authorised officer who has procured a sample and who considers that it should be examined must –

(a) if necessary place the sample in a suitable container and seal the container;
(b) mark the sample or container;
(c) as soon as it is reasonably practicable to do so, give notice to the owner that the sample will be examined; and
(d) submit it for examination.

The code of practice[1] contains important advice to authorised officers on all of the above matters and others relating to sampling.

1 Code of Practice No 7: Sampling and Analysis or Examination (HMSO ISBN 0 11 321351 4).

Give One Part to the Owner

17.4.5 Where a sample is divided into three parts, one part must be 'given to the owner' – see **17.4.3** (c) (iii) above. Under former legislation[1] the requirement was to give one part of the sample to the vendor. The new requirement has given rise to difficulties for authorised officers because there is doubt whether a part given to, for example, a store manager of a corporate body is in fact given to the owner. The view of the Ministry of Agriculture, Fisheries and Food appears to be that the manager, being a senior person in the employ of the owner, would be deemed to be the owner for this purpose. Another view is that strict compliance can only be achieved if the part of the sample is given to the company at its registered office. It will be unfortunate if the point has to be settled by the courts instead of by a simple amendment of the regulations.

1 Food Act 1984, Part I of Sch 7.

Give Notice

17.4.6 The notice to be given to the owner of the food by the authorised officer under the sampling procedures discussed above must be in writing.[1] No form for such a notice has yet been prescribed.

1 Food Safety Act 1990, s 49 (1) (b).

Certificates of Analysis or Examination

17.4.7 The owner of a food which has been sampled or examined has no statutory right to request a copy of the certificate of analysis or examination as he had under former legislation. However, the code of practice[1] recommends that this facility should be available. Also under the Food Act 1984 and its predecessors, it was necessary to serve a copy of such a certificate with the summons in any proceedings where a sample had been taken. This requirement has not been re-enacted in the Food Safety Act 1990 but a certificate should still be served in accordance with the advance information rules.[2]

1 Code of Practice No 7: Sampling and Analysis or Examination (HMSO ISBN 0 11 321351 4).
2 Magistrates' Courts (Advance Information) Rules 1985/601.

17.5 Metrological Controls

Weighing and Measuring Equipment and Quantity Controls

17.5.1 Enforcement of statutory requirements as to weighing and measuring equipment and the sale of food by weight or measure is the responsibility of local weights and measures authorities.[1] They are the metropolitan district councils, county councils, London boroughs, the Common Council of the City of London, the Inner and Middle Temples and the Council of the Isles of Scilly in England; county councils in Wales; and regional or islands councils in Scotland. The officers responsible for enforcement are inspectors of weights and measures under the direction of chief inspectors of weights and measures who may have a deputy. These officers are generally known as trading standards officers and are usually the same officers as those responsible for the enforcement of those functions under the Food Safety Act 1990 as are allocated to county councils under the Food Safety Act 1990 – see Chapter 2.

1 Weights and Measures Act 1985, s 69.

Enforcement Powers

17.5.2 General powers of inspection, entry, seizure and prosecution and the offence of obstruction are similar to those in the Food Safety Act 1990.[1]

1 Weights and Measures Act 1985, ss 79, 80, and 81.

17.6 Conclusion

17.6.1 The difficulties created by the Food Safety (Sampling and Qualifications) Regulations 1990 discussed above, and others not directly relevant to this book, seriously detract from the proper enforcement of food law and place the interests of food businesses at a disadvantage. Put simply, the regulations do not accord with natural justice.

Chapter 18

Prosecutions

18.1 The Power to Prosecute

Official Prosecutions

18.1.1 Any enforcement authority, that is a food authority by whom provisions of the Food Safety Act 1990 or any regulations or orders are to be enforced, may institute proceedings under any provisions of the Act or any regulations or orders made or continued in force under it for any contravention thereof.[1] Under the Weights and Measures Act 1985, proceedings may be instituted by or on behalf of a local weights and measures authority or the chief officer of police for a police area.[2] In Scotland the decision on whether to prosecute or not lies with the Procurator Fiscal. In England and Wales food authorities may take prosecutions themselves.

The ministers may institute proceedings for the breach of any provisions of the Act, regulations or orders for which they are the enforcement authority[3] and the ministers may also take over the conduct of any proceedings which have been instituted by some other person.[1]

1 Food Safety Act 1990, s 6 (5).
2 Weights and Measures Act 1985, s 83.
3 Food Safety Act 1990, s 6 (3) and (4).

Private Prosecutions

18.1.2 Any person other than an enforcement authority may institute proceedings under the Food Safety Act 1990 but not the Weights and Measures Act 1985. This is rarely done because private citizens do not have the powers of entry and enforcement enjoyed by authorised officers of food authorities, and the necessary evidence to secure a conviction is difficult to come by. It is better for a private citizen who has been injured by or who has suffered some loss through a food product to bring civil proceedings – see Chapter 3.

The Criteria for Prosecution

18.1.3 Code of Practice No 2[1] suggests that before deciding whether a prosecution should be taken or, in Scotland, whether the matter should be referred to the Procurator Fiscal, food authorities should consider a number of factors which may include:

(a) the seriousness of the alleged offence;
(b) the previous history of the party concerned;
(c) the likelihood of the defendant being able to establish a due diligence defence (see Chapter 19);
(d) the ability of any important witnesses and their willingness to cooperate;
(e) the willingness of the party to prevent a recurrence of the problem;
(f) the probable public benefit of a prosecution and the importance of the case – eg whether it might establish legal precedent in other companies or in other geographical areas;
(g) whether other action, such as issuing a formal caution in accordance with Home Office Circular 59/1990 (except in Scotland) or an improvement notice, or imposing a prohibition, would be more appropriate or effective (it is possible in exceptional circumstances to prosecute as well as issuing a notice; failure to comply with a notice would be an additional offence);
(h) any explanation offered by the affected company.

The Code also suggests that for England and Wales the Code for Crown Prosecutors is a valuable reference. It deals with considerations similar to those which food authorities have to address. Copies can be obtained from the Crown Prosecution Service.

In recent years there has been concern about the habit of some district councils to prosecute almost automatically whenever a case of food containing foreign bodies was reported to them. The cases were sometimes brought without any enquiries to the home authority as to the record of the company concerned, and often for relatively harmless non-toxic foreign bodies such as milk bottle tops, straws and the like. The courts have frowned on the prosecution of trivial cases[2] but it is likely that the availability of a defence of all due diligence etc (see Chapter 19) and the recommendations in the relevant code of practice[1] will reduce the numbers of such cases.

1 Code of Practice No 2: Legal Matters, Part C, (HMSO ISBN 0 11 321353 0).
2 *Smedleys Ltd v Breed* [1974] AC 839, [1974) 2 All ER 21, HL; *Sunblest Bakeries Ltd v Andrews* (1985) unreported.

The Decision to Prosecute

18.1.4 When the Food Safety Bill was being considered some consultees suggested that the decision to prosecute and the conduct of proceedings should be carried out by an independent agency such as the Crown Prosecution Service. It was believed to be undesirable that prosecution should be by the authority which carried out the investigation and that the considerations which resulted in the creation of the Crown Prosecution Service were equally valid in food law. In the event, probably for practical rather than theoretical reasons, no changes were made to the established system.

In most county councils and some large district councils the power to institute proceedings is delegated by the council to the authority's solicitor or the chief trading standards or environmental health officer. In other authorities the decision is made by a committee, sometimes in open session with the press present. This is a deplorable practice because it may lead to damaging publicity against the food business concerned before the case is heard by a court. It passes understanding why the law officers of the Crown have not put a stop to this practice.

The Conduct of Proceedings

18.1.5 The prosecution of serious food law offences is a demanding task. The law is complex and the maximum penalties are substantial. It requires a thorough knowledge of the law and relevant judicial precedents and a high degree of advocacy skill. Any member or officer of a local authority who is authorised by that authority to prosecute or defend on their behalf, or to appear on their behalf in proceedings before a magistrates' court, may do so notwithstanding that he is not a solicitor holding a current practising certificate.[1] Some years ago it was common for senior trading standards officers of large authorities to conduct food law cases and some of them became formidable advocates. A few survive, but nowadays it is usual for the prosecuting solicitor for the authority or, in serious cases, counsel to conduct the proceedings. Bearing in mind that many offences against food law are triable either way (see **18.1.8** below) this is a welcome development.

In the defence of proceedings it is important for food companies to ensure that their interests are represented by a solicitor or counsel experienced in food law and with the necessary skills to comprehend complex legal and scientific issues. Large retailers and manufacturers often employ lawyers whose close familiarity with food law makes them a match for experienced prosecutors. Conscious of the growing importance of food law some firms of solicitors now specialise in the subject and there are a number of counsel who have established a reputation in food law work.

1 Local Government Act 1972, s 223.

Commencement of Proceedings

18.1.6 All proceedings for an offence against the Food Safety Act 1990 (including regulations and orders), except for obstruction, must be begun within three years from the commission of the offence or one year from its discovery by the prosecutor, whichever is the earlier.[1] It is fatal to the proceedings if these time limits are not observed. Unreasonable delay in bringing proceedings may be an abuse of process notwithstanding the fact that they have been commenced within the statutory time limits.[2] Code of Practice No 2[3] recommends that prosecutions should be brought without unnecessary delay.

1 Food Safety Act 1990, s 34.
2 *Daventry District Council v Olins* (1990) 154 JP 478, DC; *R v Telford Justices, ex p Badhan* [1991] 2 QB 78, [1991] 2 All ER 854.
3 Code of Practice No 2: Legal Matters, Part C, (HMSO ISBN 0 11 321353 0).

First Step in a Prosecution

18.1.7 The first step in legal proceedings is to lay an information and then to issue a summons. These are standard procedures to be carried out in cooperation with the clerk to the justices. In food law, it is important to bear in mind that an information may be bad for duplicity or uncertainty. For example, to cite that the offence was 'sale to the prejudice of the purchaser of a food which was not of the nature or substance or quality' under s 14 of the Food Safety Act 1990 would be bad for duplicity because it cited more than one offence. To allege an offence as to misleading presentation of a food without stating whether the offence was against reg 4 of the Food Labelling Regulations 1984 or s 15 (3) of the Food Safety Act 1990 would be bad for uncertainty because the defendant would not know which enactment he was alleged to have contravened.

For the procedures for laying an information and the issue of a summons see the Magistrates' Courts Rules 1981/552.

The Mode of Trial

18.1.8 All but one of the offences against the Food Safety Act 1990 are triable either way, that is summarily (by the magistrates) or by the Crown Court on indictment. All offences against food regulations are triable as stated in the regulations, but most are triable summarily only.

The procedure in a case which is triable either way[1] is that the charge is read to the defendant and then he and the prosecutor are asked whether they elect for summary trial or trial in the Crown Court. The court then decides which mode of trial it thinks appropriate. In reaching its decision the court must have regard to the nature of the case, whether the circumstances make the offence one of a serious character and whether summary powers, including the prescribed maximum penalties, would be adequate. In addition, the court may have regard to other matters but not the defendant's antecedents.[2]

The decision as to whether to apply to the magistrates for trial on indictment is often difficult. The advantages of summary trial are that the case will be heard quickly and the maximum penalties which may be imposed are lighter. However, if the case is an important one, whether perceived as such by the prosecution or the defence, and the point at issue is of significant importance, perhaps having an effect beyond the case itself, then serious consideration should be given to an application for trial on indictment. If an application for such trial is successful, then it must be accepted that it will be long delayed, costs will increase considerably and penalties may be higher.

1 Magistrates' Courts Act 1980, ss 18–23.
2 *R v Colchester Justices, ex p North Essex Building Co Ltd* [1977] 3 All ER 567, [1977] 1 WLR 1109.

Maximum Penalties

18.1.9 Maximum fines for food safety offences were dramatically increased from 1 January 1991 from £2,000 to £20,000.[1] All other maximum fines are expressed as an appropriate level on the standard scale. That scale is amended from time to time to keep fines in line with inflation. The scale at the time of writing is:[2]

Level on the scale	*Amount of fine*
1	£200
2	£500
3	£1,000
4	£2,500
5	£5,000

The maximum fine on summary conviction for offences against s 7 (rendering food injurious to health), s 8 (food safety requirement) and s 14 (food not of the nature or substance or quality demanded) of the Food Safety Act 1990 is £20,000. The maximum fine on summary conviction for most other offences against the Act is the statutory maximum, ie £5,000. In addition to the above fines, magistrates may impose a prison sentence of not more than six months instead of or in addition to a fine. The maximum period of a prison sentence for an offence of obstruction of three months.

The maximum penalties on conviction on indictment are an unlimited fine and/ or imprisonment for up to two years.

Maximum penalties for offences against regulations or orders are laid down in the regulations or orders.

1 Food Safety Act 1990, s 35 (3).
2 Criminal Justice Act 1991, s 17 (1) – in force from a date to be appointed.

Formal Cautions

18.1.10 It has been the practice for many years for enforcement authorities to issue a letter of caution where it was felt that an offence had been committed but, for whatever reason, a prosecution was inappropriate. Such cautions had no status in law and could not be cited in the event of any subsequent proceedings. In 1985 the Home Office drew the attention of prosecuting local authorities to a more formal system of cautioning and this was updated and improved in 1990.[1] In cases where there is evidence which would justify a prosecution, the defendant may be asked to admit the offence and agree to being cautioned in writing. If he does so, that caution would form part of his antecedents in the event of any subsequent prosecutions. This system is being increasingly used by food authorities. It would, of course, be most unwise for a person to accept this procedure unless he was quite certain that he had committed an offence and would be convicted if he were prosecuted.

1 Home Office Circular 59/1990.

Liability of Bodies Corporate

18.1.11 A body corporate is a company formed and registered under or continued in being by the Companies Act 1985.[1] In the language of the EC it is a 'legal person'. A company can be prosecuted and punished in its own name but where an offence which has been committed by a body corporate is proved to have been committed with the consent or connivance of, or to be attributable to, any neglect on the part of any director, manager, secretary or other similar officer, or any person who was purporting to act in any such capacity, that person as well as the body corporate must be deemed to be guilty of that offence, and is liable to be prosecuted and punished accordingly.

1 Companies Act 1985, ss 1 and 735 (1).

Act or Default of Another Person

18.1.12 Where the commission of an offence by any person is due to the act or default of some other person, that other person is guilty of the offence; and a person may be charged with and convicted of the offence whether or not proceedings are taken against the first-mentioned person. This is what is known as the 'by-pass procedure' by which a food authority may bring proceedings against a supplier, agent or employee in addition to or instead of the principal offender. It is important that investigating officers should identify and interview all potential 'other persons' in every case so that the food authority is in a position to bring proceedings against them if necessary.

18.2 Evidence

Investigation Techniques

18.2.1 Although many food law cases are simple enough and require no great investigative skills to secure a conviction, others are very complex. The investigating officer may have to trace a food product from source, through a chain of distribution to the point where the primary offence has been committed, establishing at each stage the necessary evidence and assessing in each case the culpability of every company, director and manager involved. In addition to a thorough knowledge of food law he must know the law of evidence, understand the scientific and technical matters in the case, and present a clear, unambiguous and rational report of his findings. These tasks require a meticulous approach, an enquiring mind, tenacity and the ability to gain the confidence of persons being interviewed. Not all authorised officers have these qualities no matter how excellent they may be in other aspects of their work. Food cases may be lost because of failure to adduce the necessary evidence or to follow correct procedures.

Witness Statements

18.2.2 It is common practice in food cases to secure the evidence required from witnesses in written statements. Such statements may be from a person making a complaint about a food product, from any other witnesses required by the prosecution, from the investigating officers and/or their assistants, from the accused, or from any directors or managers concerned with the case. Such statements must be taken in the proper form and in accordance with the correct procedures.[1] The withholding of any evidence favourable to the accused may be fatal to the case.[2]

1 Criminal Justice Act 1967, s 9.
2 *R v Leyland Justices, ex p Hawthorn* [1979] QB 283, [1979] 1 All ER 209.

Documentary Evidence

18.2.3 In food cases documents, such as recipes, production records, staff instructions, invoices and correspondence, are often necessary as evidence. All such evidence must be obtained and used fairly. A court may refuse to allow evidence on which the prosecution proposes to rely if it appears to the court that, having regard to all the circumstances, including circumstances in which the evidence was obtained, the admission of the evidence would have such an adverse effect on the proceedings that it ought not to be admitted.[1] Letters written to the accused by the investigating officer should include a caution (see **18.2.6** below). If they do not, the accused may submit in any subsequent proceedings that he replied believing that no prosecution was intended and the correspondence may be inadmissible.

However, a written statement made by a person to an investigating officer is admissible in criminal proceedings as evidence of any fact of which direct oral evidence by him would be admissible if either:

(a) the person who made the statement in the document is dead, or by reason of his bodily or mental condition unfit to attend as a witness; or
(b) the person who made the statement is outside the United Kingdom and it is not reasonably practicable to secure his attendance; or
(c) all reasonable steps have been taken to find the person who made the statement, but he cannot be found.[2]

Thus a statement made by, say, a manager as to facts relating to an offence may be admissible in evidence in the above circumstances.

Similarly, business documents may be admissible as evidence of any fact of which direct oral evidence would be admissible if the following conditions are satisfied:

(a) the document was created or received by a person in the course of a trade or business, profession or other occupation, or as the holder of a paid or unpaid office; and
(b) the information contained in the document was supplied by a person (whether or not the maker of the statement) who had, or may reasonably be supposed to have had, personal knowledge of the matters dealt with.[3]

1 Police and Criminal Evidence Act 1984, s 78.
2 Criminal Justice Act 1988, s 23.
3 Criminal Justice Act 1988, s 24.

Computer Records

18.2.4 Records kept by food authorities and food companies are increasingly held in computer systems which may have to be presented as evidence. Documents produced by a computer are not admissible as evidence of any fact stated therein unless it is shown that:

(a) there are no reasonable grounds for believing that the statement is inaccurate because of improper use of the computer;
(b) at all material times the computer was operating properly, or if not, that any respect in which it was not operating properly or was out of operation, was not such as to affect the production of the document or the accuracy of its contents; and
(c) any relevant conditions specified in rules of court as to the form and time of production of the information.[1]

Thus, an investigating officer exercising the powers described in **17.2.4** should be in a position to verify the accuracy of any computer records he wishes to admit as evidence by an additional statement from the person in charge of the computer as to its efficient performance at the relevant time.

1 Police and Criminal Evidence Act 1984, s 69, Sch 3, para 11.

Expert Reports

18.2.5 In addition to the admissibility of certificates of analysis or examination of food, for which special provision is made in the Food Safety Act 1990 – see **17.3.5**, the prosecutor in a food case may wish to submit reports of other expert witnesses to the court. Similarly, a food company may wish to use reports from experts on food science or technology in defence of proceedings. In such cases an expert report is admissible whether or not the person making it attends to give oral evidence, but only with the leave of the court. For the purpose of determining whether to give leave, the court is required to have regard to the report, the reason why the person making the report shall not give oral evidence, any risks of unfairness to the accused and any other relevant circumstances. An expert report when admitted is evidence of any fact or opinion of which the person making it could have given oral evidence. For these purposes, an 'expert report' means a written report by a person dealing wholly or mainly with matters of which he is (or would if living be) qualified to give expert evidence.[1]

For a discussion on the use of 'experts' in establishing a defence of all due diligence see Chapter 19.

1 Criminal Justice Act 1988, s 30.

Interviews with and Statements by Suspected Persons

18.2.6 Persons other than police officers who are charged with the duty of investigating offences or charging offenders must, in the discharge of that duty, have regard to any relevant provision of a code of practice made under the Police and Criminal Evidence Act 1984.[1] Failure to comply with the provisions of any relevant code of practice would not render the officer liable to civil or criminal proceedings[2] but any evidence so obtained may be inadmissible. Failure to give a caution may not be fatal to any subsequent proceedings. The test is whether any unfairness to the defendant has resulted from such failure.[3] In food investigations, it is submitted, there is always a risk that failure to give a caution may be unfair to a defendant and the caution should always be given.

The relevant code of practice for authorised officers of food authorities or weights and measures authorities is Code C and its provisions are summarised below:

1. A person of whom there are grounds to suspect an offence must be cautioned before any questions are put to him for the purpose of obtaining evidence which may be given to a court in a prosecution. He need not be cautioned if questions are put to him to establish his identity.
2. The caution should be as follows: 'You do not have to say anything unless you wish to do so, but what you do say may be given in evidence'. Minor deviations do not constitute a breach of this requirement so long as the sense of the caution is preserved.
3. If a break in questioning occurs, upon resumption the interviewing officer must be sure that the person being questioned is aware that he is still under a caution. If there is any doubt, the caution should be given again in full when the interview resumes.
4. A record of when a caution was given should be made by the interviewing officer.
5. If a person does not understand what the caution means, the officer who has given it should go on to explain it in his own words. Should the significance of the caution be unclear, the interviewing officer should explain that it is administered in accordance with the general principle of English law that a person need not answer any question which might tend to incriminate him.
6. An accurate record must be made of each interview.
7. No officer may use oppression to obtain answers to his questions.
8. A record must be made of the place of the interview, the time it begins and ends, the time the record is made (if different), any breaks in the interview, and the names of all those present.
9. The record must be made during the course of the interview, unless in the investigating officer's view this would not be practicable or would interfere with the conduct of the interview. It should consitute a verbatim record of what has been said or, failing this, an account of the interview which adequately summarises it.
10. If an interview record is not made during the course of the interview it must be made as soon as practicable afterwards.
11. Written interview records should be timed and signed by the maker and if the interview record is not completed in the course of the interview the reason should be recorded.
12. [*This rule applies only to interviews taking place at a police station*].

13. Should a person under caution wish to make a statement he must be invited to write down himself what he wants to say, without prompting. He should sign a declaration to this effect before making the statement.
14. If a person says he would like someone to write the statement for him, this is acceptable so long as a declaration to this effect is signed, before the making of the statement.
15. Where someone other than the suspect writes the statement he must take down the exact words spoken by the person making it. Any questions necessary to make it intelligible must be recorded.
16. The forms of the two declarations in paragraphs 13 and 14 are laid down in the code of practice, together with a third declaration to be signed after the statement has been written by someone else.
17. The codes detail the procedure to be used when the identification of persons is necessary. The identification of photographs is covered by Code D.

All interviews with suspects by authorised officers must be carried out in accordance with the foregoing provisions. Officers should bear in mind the possibility that other persons by whose act or default an offence may have been committed, or directors and managers etc of the suspected company, may also be liable to prosecution and should therefore be interviewed under the above procedures.

The Police and Criminal Evidence Act 1984 fundamentally changed the law of evidence in criminal cases. It is important that all investigating officers should be familiar with its requirements and follow them precisely.

1 Police and Criminal Evidence Act 1984, s 67 (9).
2 Police and Criminal Evidence Act 1984, s 67 (10.
3 *Pennycuik v Lowe* (1991) Times, 13 December.

Tape-recorded Interviews

18.2.7 Code B issued under the Police and Criminal Evidence Act 1984 provides for the tape recording of interviews with suspects. Such tape recording is being increasingly used by authorised officers under the Food Safety Act 1990 who are obliged to have regard to the code of practice and the practice direction thereon.[1]

1 *Practice Note (Criminal Evidence) (Tape Recording of Police Interviews)* [1989] 2 All ER 415, [1989] 1 WLR 631.

The Authorised Officer's Report

18.2.8 At the conclusion of his investigation the authorised officer should prepare a written report on the case for submission to the chief officer of the department or the officer authorised to receive such reports and to decide on proceedings. There is no prescribed form for such report and practices vary between food authorities. The report should contain all the relevant facts as to the identity of the company and/or persons alleged to have committed the offence, previous convictions and other facts, followed by all the admissible evidence in logical and chronological sequence. Opinions and recommendations as to appropriate action are helpful but should be separate from the main body of the report.

Chapter 19

Defences

19.1 Strict Liability

Introduction

19.1.1 For many years it has been a cardinal principle of English criminal law that 'a man's deed does not make him guilty unless his mind be guilty'. Offences where this principle persists are easily identifiable because they contain the words 'wilfully', 'knowingly', 'intentionally', 'permitting' or 'causing' and the like. In all such cases it is necessary for the prosecutor to prove, beyond reasonable doubt, that the offender had a guilty mind, that is that he knew that the act or omission he is accused of was unlawful. Such offences continue to appear in modern trading law statutes[1] but generally the above words are missing. Where that is so, the offence is an absolute one, that is the mere commission of the unlawful act or the failure to do something required by law is sufficient for conviction. It is not necessary for the prosecutor to prove that the offender had a guilty mind. Most of the principal offences in the Food Safety Act 1990 and the Weights and Measures Act 1985 are of this type. It was established more than a hundred years ago[2] that an offence against s 6 of the Sale of Food and Drugs Act 1875 (similar to s 14 of the Food Safety Act 1990) had been committed even though the accused did not know that the food was not of the nature or substance or quality demanded.

Offences of absolute liability are employed in trading legislation because it would be near impossible to secure a conviction if the prosecutor was obliged to prove guilty intent in every case. It would merely be necessary for the accused to acknowledge the facts but then state that he was unaware of the offence and the court would be obliged to acquit him. Conscious, however, that absolute liability could bear harshly on traders, Parliament introduced a series of statutory defences which would, subject to proof that the criteria in each case had been fulfilled, enable a court to acquit a trader even though the offence had been committed. Statutory defences have evolved over the years and the Food Safety Act 1990 brought those relating to food offences up-to-date.

1 See, for example, the offence of contravening a prohibition order under s 11 (5) of the Food Safety Act 1990.
2 *Betts v Armstead* (1888) 20 QBD 771.

The Warranty Defence

19.1.2 The Food Act 1984[1] and its predecessors contained what was generally called 'the warranty defence'. Put simply, it provided that if a trader was in possession of a warranty that the goods he had bought for sale complied with all relevant legislation he could, subject to certain requirements, be absolved of any responsibility for the food he was selling. It was open to the prosecution to charge the warrantor for giving a false warranty or, as appropriate, charge the warrantor with being the person by whose act or default that offence was committed, but the principal offender, usually the retailer, could rely on the warranty. The pattern of food law enforcement was influenced over many years by the concept of warranty but it had a serious defect. An importer could rely on a warranty given by an overseas supplier but, because that supplier was outside the jurisdiction of the United Kingdom courts, he could not be prosecuted. Thus importers of food enjoyed a degree of protection against prosecution which was not available to United Kingdom manufacturers. For this reason and others the defence of warranty was not re-enacted in the Food Safety Act 1990 although a guarantee from a supplier may continue to have a role in the new defence created by that Act.

1 Food Act 1984, s 102.

19.2 The Due Diligence Defence

Due Diligence

19.2.1 Some 30 years ago a defence of 'all reasonable precautions and all due diligence' began to appear in trading law statutes. Examples may be found in s 24 of the Trade Descriptions Act 1968, s 168 of the Consumer Credit Act 1974, s 34 of the Weights and Measures Act 1985 and s 39 of the Consumer Protection Act 1987. There was such a defence in s 100 of the Food Act 1984 in much the same form as had existed in similar statutes of 1928, 1938 and 1955, but it was linked to proof that the offence was due to the act or default of another person and, to take advantage of it, it was necessary for the accused himself to charge that other person, unless the prosecuting authority chose to do so. It was a cumbersome defence, rarely used, and of little practicable value to food companies.

The Food Safety Act 1990[1] has introduced a defence similar to that in the statutes cited above but in modified form to take account of modern food manufacturing and distribution practice. It states:

> '. . . it shall . . . be a defence for the person charged to prove that he took all reasonable precautions and exercised all due diligence to avoid the commission of the offence by himself or by a person under his control'.

That is the basic, simple defence similar to that used in other statutes. However, the Food Safety Act 1990 makes additional provisions[2] for persons who neither prepared nor imported the food and who are accused of offences under s 8 (food not complying with the food safety requirement), s 14 (food not of the nature or substance or quality demanded) or s 15 (false or misleading labelling or advertising of food). Such a person is deemed to have established the defence given above if he proves either:

(I)

(a) 'that the commission of the offence was due to an act or default of another person who was not under his control, or to reliance on information supplied by such a person;

(b) that he carried out all such checks of the food in question as were reasonable in all the circumstances, or that it was reasonable in all the circumstances for him to rely on checks carried out by the person who supplied the food to him; and
(c) that he did not know and had no reason to suspect at the time of the commission of the alleged offence that his act or omission would amount to an offence under the relevant provision'.

or:

(II)
(a) 'that the commission of the offence was due to an act or default of another person who was not under his control, or to reliance on information supplied by such a person;
(b) that the sale or intended sale of which the alleged offence consisted was not a sale or intended sale under his name or mark; and
(c) that he did not know, and could not reasonably have been expected to know, at the time of the commission of the alleged offence that his act or omission would amount to an offence under the relevant provision'.

1 Food Safety Act, s 21 (1).
2 Food Safety Act 1990, s 21 (2), (3) and (4).

The 'Deemed Provisions'

19.2.2 The qualifications to the simple due diligence defence set out in **19.2.1** (I) and (II) above have become known as 'the deemed provisions'. Their objective is to offer a lesser burden of proof to persons who neither prepared the food nor imported it. They are an attempt to place responsibility for food upon the persons who have the greatest influence over the final product. Put simply, the requirement is that someone, somewhere, must ensure that food is throughly checked and controlled to ensure compliance with the law.

Persons who prepare food must fulfil the whole of the due diligence defence and cannot take advantage of the deemed provisions. It is important to understand that 'preparation' includes the manufacture and any form of processing or treatment; and 'treatment' includes subjecting food to heat or cold.[1] Thus 'preparers' includes manufacturers, processors, caterers and retailers who, for example, bake bread and flour products themselves; who heat pies or other products, who chill or freeze products; or who add cream or other garnishments to food. Similarly, importers of food into the Great Britain (see **4.3**) must satisfy the whole of the defence.

However, retailers or any other class of traders other than those above can take advantage of the deemed provisions in respect of offences against ss 8, 14 or 15 of the Act. Such retailers are divided into two groups; those who sell food products under their own name or mark (own-labellers) and those who do not, ie those who sell only branded goods. Own-labellers have a higher level of responsibility than retailers of branded goods because they must be able to prove all three limbs of (I) above. Sellers of branded goods have only to prove (II) (a) and (c) which, as discussed below, is an easier task.

1 Food Safety Act 1990, s 53 (1).

The Burden of Proof

19.2.3 There is nothing in the Food Safety Act 1990, nor any other relevant statute, which *requires* a food business to have a due diligence system. It is open to a food business to simply sit back and do nothing, taking a chance on the commission of an offence and the tremendous risks that such would involve. No rational business person would take such a risk for the penalties and the notoriety attached to food law offences are such that a business may not survive the consequences of a serious offence. For all practical purposes it is essential that every food business should establish and maintain an adequate due diligence system.

The burden of proof of due diligence rests with the defendant in any proceedings. He must be able to prove to the satisfaction of a court that the control system is adequate, bearing in mind the nature of the products sold and the risks attached to them. The burden of proof is the civil law one of 'on the balance of probability'.[1]

1 *R v Carr-Briant* [1943] KB 607, [1943] 2 All ER 156, CCA; *Robertson v Watson* 1949 JC 73; *R v Jenkins* (1923) 87 JP 115, CCA; *Cant v Hartley & Sons Ltd* [1938] 2 All ER 768.

Setting up the Control System

19.2.4 The courts have looked at the defence of 'all reasonable precautions and all due diligence' many times in cases brought under weights and measures, trade descriptions and consumer safety law. There are therefore a considerable number of judicial precedents which provide valuable guidance on what may be required to establish the defence under the Food Safety Act 1990. However, there have been no such cases so far under the 1990 Act and, when considering the relevance of existing precedents, regard must be had to the fact that the courts may choose to take a sterner view of food safety cases. After all, the consequences of unsafe food being sold may be a good deal more dramatic than short weight or misleading product descriptions. The following are some general rules based on existing court judgments and a projected interpretation of the defence in food law.

(a) Positive steps to set up a system are essential. Sitting back and doing nothing can never satisfy the defence.
(b) 'All reasonable precautions' means setting up a control system. 'All due diligence' means ensuring that the system works.
(c) The word 'all' requires that if there is a precaution which can be reasonably taken then it must be taken. Unless it is, the defence cannot succeed.
(d) The word 'reasonable', on the other hand, means that the control system must be capable of dealing with a perceived risk.
(e) The extent of the control system depends on the size and resources of the business. What is required for a large corporate body may not be appropriate for a small private business.[1]
(f) The control system must be capable of identifying faults in all aspects of the company's business. Because the defence in the Food Safety Act 1990 is similar to those in other relevant statutes a common control system can be established to deal with all matters subject to that Act, plus quantity controls, price checks and general product descriptions. A non-exhaustive list of subjects to be covered may be:
    (i) hygiene of premises, equipment and staff;
    (ii) raw materials, food sources and packaging;
    (iii) production, processing, packing and storage;

(iv) recipes, product specifications;
(v) labelling and advertising;
(vi) quantity marking, weighing and measuring;
(vii) price marking, misleading indications of price;
(viii) staff training;
(ix) monitoring of customer complaints.

(g) All staff participating in the system must be given written instructions and be asked to acknowledge them in writing. If possible, their duties should be identified in their job specifications.

(h) The system must be pro-active and re-active. It must be capable of identifying and preventing risks and of correcting them if they occur.

(i) The system must be kept under review and updated as necessary.

(j) The adequacy of the control system will be judged by the nature of the products, the manufacturing, processing or retailing techniques involved, and all other relevant factors. Although there are many common matters in due diligence defences, there is no such thing as a blueprint due diligence system. Every system must be unique to the company operating it.

The advice given above is based, in part, on an analysis of judicial precedents. The following is a brief summary of some of the most important of them:

Due diligence is a question of fact in each case to be assessed by the court. The High Court will interfere if there is insufficient evidence to support a finding by a lower court – *RC Hammett Ltd v Crabb* (1931) 145 LT 638; *Rogers v Barlow & Son* (1906) 94 LT 519; *RC Hammett Ltd v LCC* (1933) 97 JP 105; *Pearce v Cullen* (1952) 96 Sol Jo 132.

Every case depends on its own circumstances and there can be no standard system of due diligence. Trade or professional rules or codes of practice may contribute to a defence but cannot fulfil the whole burden of proof – *Zawadski v Sleigh* [1975] RTR 113, DC; *Amos v Melcon (Frozen Foods) Ltd* (1985) 149 JP 712, DC.

Mitigation must not be confused with defence. A plea that the commission of the offence was due to accidental circumstances is not a defence – *Marshall v Herbert* [1963] Crim LR 506.

The defendant must have taken all reasonable precautions. What is reasonable must be determined on the facts of the case and on the size and resources of the defendant's business – *Sherratt v Gerald the American Jewellers Ltd* (1970) 114 Sol Jo 147; *Garrett v Boots Chemists Ltd* (16 July 1980, unreported); *Beckett v Kingston Bros (Butchers) Ltd* [1970] 1 QB 606, [1970] 1 All ER 715.

Reliance cannot be placed on checks carried out by overseas suppliers or by an enforcement authority – the defendant must himself carry out checks or take other positive steps to ensure that the goods he sells are lawful as appropriate – *Taylor v Lawrence Fraser (Bristol) Ltd* (1977) 121 Sol Jo 757; *Hicks v SD Sullam Ltd* (1983) unreported; *Hurley v Martinez & Co Ltd* (1990) 154 JP 821.

A blanket assurance by a supplier is not enough – a warranty from a supplier may be a first step in establishing due diligence but it does not absolve the defendant from doing something positive himself – *Riley v Webb* (1987) 151 JP 372.

The number of tests carried out must be sufficient to offer a reasonable possibility of detecting a fault and the results must be properly reported and taken into account – *Rotherham Metropolitan Borough Council v Raysun (UK) Ltd* (1988) 153 JP 37; *P & M Supplies (Essex) Ltd v Devon County Council* (1991) unreported.

Lack of expert knowledge is no defence - lack of knowledge of a product being sold is no defence and in fact increases the need to make adequate checks – *Sutton London Borough v Perry Sanger & Co Ltd* (1971) 135 JP Jo 239.

Due diligence must be exercised by those having the directing mind or will of the company – ie those having the capacity to direct the operations of the company, and the default of management staff does not necessarily deprive the company of a defence – *Tesco Supermarkets Ltd v Nattrass* [1972] AC 153, [1971] 2 All ER 127, HL; *HL Bolton (Engineering) Co Ltd v TF Graham & Sons Ltd* [1957] 1 QB 159, [1956] 3 All ER 624, CA.

Comprehensive instructions and training must be given to staff – regular checks by senior staff must be established, instructions and guidelines must be given to managers and they must be trained in their duties – *Baxters Butchers Ltd v Manley* (1985) unreported.

1 *Garrett v Boots Chemists Ltd* (16 July 1980, unreported).

Own-labellers

19.2.5 It will be noted that the deemed provisions for own-labellers include specific requirements that they should have carried out all such checks of the food in question as were reasonable in all the circumstances, or that it was reasonable in all the circumstances to rely on checks by their suppliers. This is a halfway-house requirement between a full due diligence system as required of a manufacturer or importer and that required from a retailer selling branded goods. The ratio of self-checking and reliance on a supplier's checks must depend on the confidence the retailer has on the checking system operated by the supplier. This confidence can be achieved by regular checks on the supplier's due diligence system by the retailer's own qualified staff or by agents employed by him. What it amounts to is that the supplier and retailer in combination must have carried out a full due diligence system.

Written Assurances from Suppliers

19.2.6 Consequent upon the abolition of warranties as a defence in their own right, there has been some confusion about the role of written assurances from suppliers. The courts have demonstrated a number of times that it is the duty of a food business to seek written assurances from their suppliers that the products being supplied comply with all legal requirements. Such assurances are, it is suggested, an essential first step in the establishment of a due diligence system. Such assurances are not warranties in the sense used in the Food Act 1984 and earlier statutes.

Certain retailers have been seeking very detailed written assurances giving emphasis to particular points which concern them. That is all very well but others have sought assurances which the supplier cannot fulfil. The courts are unlikely to be impressed with assurances which are beyond the competence of the supplier.

Records

19.2.7 A control system which is not written down and recorded is not capable of proof in a court no matter how comprehensive or effective it may be. The basic system should be drafted and approved by the company. In a large company such approval should be by the board of directors. In a small private company it should be approved by the proprietor or partners. The checks made under the system should be recorded and signed by the staff involved. Defects should be reported to senior management in writing and the action taken to rectify them should be similarly written down. The objective is to satisfy an investigating officer that an adequate system is in operation and the defence could be established if the matter should be taken to court. If the matter does result in a prosecution, then the system must be able to withstand cross-examination by the prosecutor and satisfy the court, as a matter of fact, that the defence is made out.

Hazard Analysis

19.2.8 This is a system known as HACCP (Hazard Analysis Critical Control Points) which has been rapidly developed in the food industry as a result of the Food Safety Act 1990. The system involves the identification and weighting of risks and matching them to appropriate control measures. It applies to all aspects of food mnaufacture and distribution and has been incorporated in two codes of practice[1] for the guidance of enforcement officers. The codes establish scoring systems by which the hazards created by different factors are assessed and the frequency and detail of inspection is based. Enforcement officers are required to arrange their inspections in accordance with the codes and it is logical for food businesses to use the criteria in them to ensure that the scope and effectiveness of their due diligence systems is based on the same criteria. It is therefore recommended that all food businesses would benefit from having copies of the codes.

The criteria used in assessing the risk presented by food businesses include the following:

Food Standards (ie composition, quality, raw materials, additives, processing, packaging materials, labelling, advertising and presentation):

1. the effectiveness of the business's quality assurance systems (if any);
2. the likelihood of the business failing to meet food standards (for example, any history of past failures);
3. the type of business;
4. the type of processes operated on the premises (if any);
5. the number and volume of production lines (where applicable);
6. the international, national or local importance of the business;

Food Hygiene (ie, all matters concerning food safety and general hygiene) – criteria given in ascending order of importance.

Type of Food and Method of Handling:

1. day-to-day retail handling of foods other than high risk, such as fruit, vegetables, canned and other ambient shelf stable products;

2. day-to-day handling of prepacked high risk foods; and preparation (including cooking) of high risk foods in establishments supplying less than 20 consumers each day; wholesalers and distributors;
3. day-to-day handling or preparation (including cooking) of open high risk foods, for example, restaurants, staff canteens, public houses, takeaways, sandwich preparation, delicatessens supplying more than 20 consumers each day; production/manufacture including packing of foods other than high risk; wholesalers of high risk foods;
4. production/manufacture of high risk foods such as cooked meat and poultry, dairy products.

High risk foods may be regarded as those foods that support the growth of micro-organisms, and/or are intended for consumption without further treatment that could destroy pathogenic micro-organisms or their toxins.

Method of Processing:

The risk assessed by the above criteria is increased by certain food processes where the potential hazard is greater. The following should be included in this category:

* manufacture of cooked and chilled food; ie cooked and prepared meals or foods which may be eaten cold or after re-heating;
* thermal processing of low acid products;
* aseptic packing of low acid products;
* vacuum packing including sous-vide but excluding raw and unprocessed meats and dried foods;
* small-scale production of cooked meat products eg certain retailers including butchers.

Consumers at Risk:

The number of consumers likely to be put at risk if there is a failure of food hygiene and safety procedures is taken into account based on the following criteria:

* very few – premises supplying less than 20 consumers each day;
* few – premises supplying essentially local trade, for example, high street or corner shops and local restaurants;
* intermediate – includes rather larger businesses whose trade extends outside of the town or district, for example, hypermarkets, small local manufacturers;
* substantial – larger manufacturers whose products are distributed nationally or internationally.

Vulnerable Groups:

The risk should be considered to be yet greater if the premises serves vulnerable groups eg, the sick, the elderly, children under five years, the immuno-compromised.

Note: The above notes are intended as a guide to the principles of HACCP only. There are a number of excellent publications on this important subject giving more detailed guidance.[2]

1 Code of Practice No 8: Food Standards Inspections (HMSO ISBN 0 11 321466 9) and Code of Practice No 9: Food Hygiene Inspections (HMSO ISBN 0 11 321467 7).

2 *Guidelines for the Establishment of HACCP – Technical Manual 19* (1987) Campden FPRA; *Health*

and Safety in Kitchens and Other Food Preparation Areas – Health and Safety (G) 55–11–1190; *Catering Safety – Food Preparation Machinery* – Health and Safety (G) 35; *Safety in Meat Preparation – guidance for butchers* – Health and Safety (G) 45 (HSE).

Quality Systems

19.2.9 Quality assurance is relevant to the setting up of a due diligence system in that it establishes the correct procedures for management and control of manufacturing, processing and selling systems. However, adherence to quality systems does not of itself establish due diligence because it does not directly address the legality of products. Nevertheless, company procedures in accordance with BS 5750 or its international equivalents[1] may contribute substantially to the acceptance of a due diligence system by enforcement officers and the courts.

1 British Standard BS 5750 (1987); BS 7229 (1990); BSI Handbook 22 (1987); International Standards Organisation ISO 9000 series; Quality Systems for the Food and Drink Industries.

Setting Standards in a Due Diligence System

19.2.10 To achieve the credibility of a due diligence system it is desirable to have scientific and technical parity with the enforcement authorities. Those authorities enjoy the services of professionally qualified environmental health officers and trading standards officers, and public analysts and food examiners holding statutory qualifications (see **17.3.3**). It is therefore suggested that in-house technical control staff and laboratories should, if they are not under the direct supervision of a suitably qualified person, be subject to audit by outside experts. There are a number of companies offering quality systems and due diligence control services, who employ food analysts and food examiners holding the statutory qualifications, and experienced environmental health and trading standards officers holding the necessary professional qualifications. Consultant laboratories should have obtained independent accreditation (eg under the NAMAS scheme) and operate quality systems to BS 5750. They should also be covered by professional indemnity insurance.

It should also be borne in mind that persons involved in the operation of due diligence systems by whose act or default an offence is committed may be subject to prosecution.[1]

1 Food Safety Act 1990, s 20.

Other Guidance on Due Diligence Systems

19.2.11 There are a number of sources of guidance on due diligence systems. Foremost among them is the *Guide to Good Manufacturing Practice* of the Institute of Food Science and Technology, now in its third edition, which takes account of recent changes in the law. It is highly recommended.[1]

Also recommended are *Guidelines on the Statutory Defence of Due Diligence* prepared by the National Consumer Council, LACOTS, the Institution of Environmental Health officers, the National Farmers Union, the Retail Consortium and the Food & Drink Federation. It has been supported by the Parliamentary Secretary to the Ministry of Agriculture, Fisheries and Food. Copies may be obtained from any of the above organisations.

1 *A Guide to Good Manufacturing Practice*, IFST, 5 Cambridge Court, 210 Shephard's Bush Road, London W6 7NL: price £25.

19.3 Innocent Publication

19.3.1 The Food Safety Act 1990 repeats the now common defence for persons who innocently publish advertisements.[1] It provides that it is a defence for the defendant to prove that he is a person whose business it is to publish or arrange for the publication of advertisements and that he received the offending advertisement in the ordinary course of his business and did not know, and had no reason to suspect, that its publication would amount to an offence.

1 Food Safety Act 1990, s 22.

Publishers/agents

19.3.2 While a publisher of advertisements who does not contribute to the form or content of an advertisement would be protected by this defence, the same could not necessarily be said of an advertising agency involved in the design of advertisements. It is necessary for such agencies to keep a close record through contact reports with the advertiser as to the final approval of an advertisement. Any feature of an advertisement published without the approval of the advertiser could create criminal liability for the agency.

19.4 Defences in Regulations

Due Diligence and Publishing Defence

19.4.1 Most of the regulations made or continued in operation by the Food Safety Act 1990 have the defences of due diligence and innocent publication contained in the Act applied to them. Thus a due diligence system must cover all legal requirements imposed by regulations.

Special Defences

19.4.2 Certain regulations, however, offer additional defences for particular requirements created by them. There is, for example, a modified due diligence defence in relation to meat content requirements[1] and special defences that foods not marked with certain particulars would be so marked before sale.[1,2] All food businesses whose products are subject to specific regulations should ascertain whether such defences are provided and take steps to incorporate them in their general due diligence systems.

1 Meat Products and Spreadable Fish Products Regulations 1984/1566, reg 16.
2 Food Labelling Regulations 1984/1305, reg 42.

Appendix A

Digest of Food Legislation

Statutes applicable in whole or in part to food

Food Safety Act 1990 (c 16) – food safety, hygiene, quality, labelling, advertising, processes and contaminants, food packaging materials, control of food premises.

Food and Environment Protection Act 1985 (c 48) – emergency orders as to food contamination, control of pesticides.

Consumer Protection Act 1987 (c 43) – strict product liability, food imitations and misleading indications of price.

Scotch Whisky Act 1988 (c 22) – injunctive action re Scotch whisky which does not conform to a definition.

European Communities Act 1972 (c 68) – implementation of certain EC instruments.

Slaughterhouses Act 1974 (c 3) – hygiene in slaughterhouses.

Agricultural Produce (Grading and Marking) Act 1928 (18 & 19 Geo 5, c 19) – grading and marking of agricultural produce.

Agriculture and Horticulture Act 1964 (c 28) – grading of horticultural produce.

Horticultural Produce Act 1986 (c 20) – power of authorised officers as to produce-grading.

Medicines Act 1968 (c 67) – medicinal claims, distinction between medicines and foods, control of medicinal feeding stuffs for animals likely to affect foods for human consumption.

Weights and Measures Act 1985 (c 72) – control of weighing and measuring equipment; sale of food by quantity.

Weights and Measures &c Act 1976 (c 77) – orders for the alleviation of food shortages.

Trade Descriptions Act 1968 (c 29) – general false trade descriptions including false indications of origin.

Prices Act 1974 (c 24) – Orders requring giving of price information.

Prices Act 1975 (c 32) – amendment of Prices Act 1974 and regulation of prices.

Note: Most of the above statutes have been amended a number of times. For the full annotated text of each statute see *Butterworths Law of Food and Drugs*, *O'Keefe's Law of Trade Descriptions* and *O'Keefe's Law of Weights and Measures* (Butterworths).

Orders and Regulations applicable in whole or in part to food

Food Safety, Hygiene, Quality, Labelling etc

Public Health (Shell-Fish Regulations) 1934/1342 as amended by SIs 1948/1120, 1982/1727 and 1990/2486.

Food Standards (General Provisions) Order 1944/42 as amended by SR&O 1944/654 and SIs 1982/1727 and 1985/67.*

Food Standards (Fish Cakes) Order 1959/589.*

Milk and Dairies (Channel Islands and South Devon Milk) Regulations 1956/919 as amended by SIs 1962/1288, 1982/1703, 1985/68 and 1990/2486.

Milk and Dairies (General) Regulations 1959/277 as amended by SIs 1962/1288, 1973/1064, 1977/171, 1979/1567, 1982/1703, 1985/68 and 1990/2486.

Ice-Cream (Heat Treatment etc) Regulations 1959/734 as amended by SIs 1962/1287, 1963/1083, 1982/1727, 1985/67 and 1990/2486.

Arsenic in Food Regulations 1959/831 as amended by SIs 1960/2261, 1962/1287, 1963/1435, 1972/1391, 1973/1052, 1973/1340, 1975/1486, 1982/1727, 1984/1304, 1985/67, 1990/2486 and 1991/1476.

Food Hygiene (Docks, Carriers etc) Regulations 1960/1602 as amended by SIs 1962/1287, 1982/1727, 1985/67, 1990/1431 and 1990/2486.

Skimmed Milk with Non-milk Fat Regulations 1960/2331 as amended by SIs 1962/1287, 1976/103, 1980/1849, 1981/1174, 1982/1727, 1985/67, 1990/607, 1990/2486 and 1991/1476.

Liquid Egg (Pasteurisation) Regulations 1963/1503 as amended by SIs 1982/1727, 1985/67 and 1990/2486.

Meat (Treatment) Regulations 1964/19 as amended by SIs 1982/1727, 1985/67 and 1990/2486.

Soft Drinks Regulations 1964/19 as amended by SIs 1969/1818, 1970/1597, 1972/1510, 1976/295, 1977/927, 1980/1849, 1982/1727, 1983/1211, 1985/67, 1990/2486 and 1991/1476.**

Eggs (Marking and Storage) Regulations 1965/1000.

Food Hygiene (Markets, Stalls and Delivery Vehicles) Regulations 1966/791 as amended by SIs 1966/1487, 1982/1727, 1985/67, 1990/1431, 1990/2486 and 1991/1343.

Mineral Hydrocarbons in Food Regulations 1966/1073 as amended by SIs 1982/1727,1985/67, 1990/2486 and 1991/1476.**

Butter Regulations 1966/1074 as amended by SIs 1973/1340, 1980/1849, 1982/1727, 1985/67, 1990/2486 and 1991/1476.**

Solvents in Food Regulations 1967/1582 as amended by SIs 1967/1939, 1980/1832, 1982/1727, 1983/1211, 1984/1304, 1985/67, 1990/2486 and 1991/1476.

Ice-Cream Regulations 1967/1866 as amended by SIs 1980/1849, 1982/1727, 1983/1211, 1985/67, 1990/607, 1990/2486 and 1991/1476.**

Margarine Regulations 1967/1867 as amended by SIs 1982/1727, 1985/67, 1990/2486 and 1991/1476.**

Cheese Regulations 1970/94 as amended by SIs 1974/1122, 1975/1486, 1976/2086, 1980/1849, 1982/1727, 1984/649, 1985/67, 1990/2486 and 1991/1476.**

Cream Regulations 1970/752 as amended by SIs 1975/1486, 1980/1849, 1982/1727, 1985/67, 1990/2486 and 1991/1476.**

Food Hygiene (General) Regulations 1970/1597 as amended by SIs 1982/1727, 1985/67, 1990/1431, 1990/2468 and 1991/1343.**

Colouring Matter in Food Regulations 1973/1340 as amended by SIs 1974/1119, 1975/1488, 1976/2086, 1978/1787, 1979/1254, 1982/1727, 1985/67, 1987/1987, 1990/2486 and 1991/1476.***

Specified Sugar Products Regulations 1976/509 as amended by SIs 1980/1834, 1980/1849, 1982/255, 1982/1727, 1985/67, 1990/2486 and 1991/1476.***

Cocoa and Chocolate Products Regulations 1976/1209 as amended by SIs 1980/1833, 1980/1849, 1982/17, 1982/1727, 1985/67, 1990/2486 and 1991/1476.**

Poultry Meat (Hygiene) Regulations 1976/1209 as amended by SIs 1979/693, 1981/1168, 1982/1727, 1985/67, 1990/2494 and 1990/2486.***

Honey Regulations 1976/1832 as amended by SIs 1980/1849, 1982/1727, 1985/67, 1990/2486 and 1991/1476.

Drinking Milk Regulations 1976/1883 as amended by SIs 1982/1703, 1985/68, 1990/2486 and 1991/1476.

Milk and Dairies (Milk Bottle Caps) (Colour) Regulations 1976/2086 as amended by SIs 1982/1703, 1985/68 and 1990/2468.

Erucic Acid in Food Regulations 1977/691 as amended by SIs 1982/264, 1982/1727, 1985/67, 1990/2486 and 1991/1476.

Fruit Juices and Fruit Nectars Regulations 1977/927 as amended by SIs 1979/1254, 1980/1849, 1982/1311, 1982/1727, 1985/67, 1990/2486 and 1991/1284.***

Condensed Milk and Dried Milk Regulations 1977/928 as amended by SIs 1980/1849, 1982/1066, 1982/1727, 1985/67, 1986/2299, 1989/1959, 1990/2486 and 1991/1476.

Antioxidants in Food Regulations 1978/105 as amended by SIs 1980/1831, 1982/1727, 1983/1211, 1984/1304, 1985/67, 1990/2486, 1991/1476 and 1991/2540.***

Coffee and Coffee Products Regulations 1978/1787 as amended by SIs 1980/1849, 1982/254, 1982/1727, 1985/67, 1987/1986, 1990/2486 and 1991/1476.

Lead in Food Regulations 1979/1254 as amended by SIs 1982/1727, 1985/67, 1985/912, 1990/2486 and 1991/1476.

Chloroform in Food Regulations 1980/36 as amended by SIs 1982/1727, 1985/67, 1990/2486 and 1991/1476.

Miscellaneous Additives in Food Regulations 1980/1834 as amended by SIs 1980/1849, 1982/14, 1982/1727, 1983/1211, 1985/67, 1990/399, 1990/2486 and 1991/1476.***

Jam and Similar Products Regulations 1981/1063 as amended by SIs 1982/1700, 1982/1727, 1983/1211, 1985/67, 1988/2112, 1989/533, 1990/2085, 1990/2486 and 1991/1476.

Grading or Horticultural Produce (Forms of Labels) Regulations 1982/387

Meat (Sterilisation and Staining) Regulations 1982/1018 as amended by SIs 1982/1727, 1984/604, 1985/67 and 1990/2486.***

Milk (Presentation) Regulations 1982/1702

Sweeteners in Food Regulations 1983/1211 as amended by SIs 1985/67, 1988/2112, 1989/533, 1989/876, 1990/2486 and 1991/1476.***

Milk-Based Drinks (Hygiene and Heat Treatment) Regulations 1983/1508 as amended by SIs 1985/67, 1986/720, 1990/2486 and 1991/1476.

Milk and Dairies (Heat Treatment of Cream) Regulations 1983/1509 as amended by SIs 1985/68, 1986/721, 1990/2486 and 1991/1476.

Poultry Meat (Water Content) Regulations 1984/1145 as amended by SI 1990/2486.

Bread and Flour Regulations 1984/1304 as amended by SIs 1985/67, 1989/533, 1989/876, 1990/399, 1990/2468 and 1991/1476.

Food Labelling Regulations 1984/1305 as amended by SIs 1985/71, 1985/67, 1987/1986, 1988/2112, 1989/768, 1989/2321, 1990/2486, 1990/2489, 1991/2488, 1991/2489 and 1991/1476.**

Meat Products and Spreadable Fish Products Regulations 1984/1566 as amended by SIs 1985/67, 1986/987, 1990/2468 and 1991/1476.

Imported Food Regulations 1984/1918 as amended by SIs 1990/2371 and 1990/2486.***

Natural Mineral Waters Regulations 1985/71 as amended by SIs 1990/2487 and 1991/1476.

Eggs (Marketing Standards) Regulations 1985/1271 as amended by SIs 1987/1184 and 1991/1386.

Caseins and Caseinates Regulations 1985/2026 as amended by SIs 1989/2321, 1990/2486 and 1991/1476.

Sea Fish (Marketing Standards) Regulations 1986/1272.

Authorised Officers (Meat Inspection) Regulations 1987/133 as amended by SI 1990/2486.

Materials and Articles in Contact with Food Regulations 1987/1523 as amended by SIs 1990/2486 and 1991/1476.

Olive Oil (Marketing Standards) Regulations 1987/1783 as amended by SI 1990/2487.

Meat Inspection Regulations 1987/2236 as amended by SIs 1990/2486, 1990/2494 and 1990/2495.***

Fresh Meat Export (Hygiene and Inspection) Regulations 1987/2237 as amended by SIs 1990/2486, 1990/2493 and 1990/2494.

Animals and Fresh Meat (Examination for Residues) Regulations 1988/848 moldified by SI 1988/849.

Animals and Fresh Meat (Hormonal Substances) Regulations 1988/849.

Importation of Milk Regulations 1988/1803 as amended by SIs 1990/2486 and 1991/1231.

Milk and Dairies (Semi-skimmed and Skimmed Milk) (Heat Treatment and Labelling) Regulations 1988/2206 as amended by SIs 1989/2382, 1990/2486, 1990/2491 and 1991/1476.

Preservatives in Food Regulations 1989/553 as amended by SIs 1989/2287, 1990/2486 and 1991/1476.***

Emulsifiers and Stabilisers in Food Regulations 1989/910 as amended by SIs 1990/2486 and 1991/1476.***

Tetrachloroethylene in Olive Oil Regulations 1989/910 as amended by SI 1990/2486.

Bovine Offal (Prohibition) Regulations 1989/2061 as amended by SI 1990/2486.

Meat and Meat Products (Hormonal Substances) Regulations 1989/1959 as amended by SI 1990/2486.

Milk (Special Designation) Regulations 1989/2383 as amended by SIs 1990/2486, 1990/2492 and 1991/1476.

Potassium Bromate (Prohibition as a Flour Improver) Regulations 1990/399.

Milk and Milk Products (Protection of Designations) Regulations 1990/607 as amended by SI 1990/2486.

Scotch Whisky Order 1990/998.

Preserved Sardines (Marketing Standards) Regulations 1990/1084 as amended by SI 1990/2486.

Spirit Drinks Regulations 1990/1179 as amended by SI 1990/2486.

Ungraded Eggs (Hygiene) Regulations 1990/1323) as amended by SI 1990/2486.

Common Agricultural Policy (Wine) Regulations 1990/1363.

Milk and Dairies and Milk (Special Designation) (Charges) Regulations 1990/1584 as amended by SI 1990/2486.

Tryptophan in Food Regulations 1990/1728 as amended by SI 1990/2486.

Food Safety (Enforcement Authority) (England and Wales) Order 1990/2462.

Food Safety (Sampling and Qualifications) Regulations 1990/2463.

Food Safety Act 1990 (Consequential Modifications) (England and Wales) Order 1990/2486.

Food Safety Act 1990 (Consequential Modifications) (No 2) (Great Britain) Order 1990/2487.

Food (Control of Irradiation) Regulations 1990/2490.

Fresh Meat and Poultry Meat (Hygiene, Inspection and Examinations for Residues) (Charges) Regulations 1990/2494.

Detention of Food (Prescribed Forms) Regulations 1990/2614.

Quick-Frozen Foodstuffs Regulations 1990/2615.

Food Safety Act 1990 (Consequential Modfications) (Scotland) Order 1990/2625 (S 220).

Food Safety (Improvement and Prohibition – Prescribed Forms) Regulations 1991/199.

Food Safety (Northern Ireland) Order 1991/762.

Food (Miscellaneous Revocations) Regulations 1991/1231.

Food Safety (Exports) Regulations 1991/1476.

Hormonal Substances (Food Sources) (Animals) Regulations 1991/1593.

Imported Food and Feedingstuffs (Safeguards against Cholera) Regulations 1991/2486.

Food Premises (Registration) Regulations 1991/2825.

* Temporarily remaining in force pending new regulations on fish products.
** Regulations due for early revision or consolidation.
*** Regulations soon to be substantially amended or replaced by virtue of EC measures.

Regulations as to Hygiene in Slaughterhouses

Slaughterhouses (Hygiene) Regulations 1977/1805 as amended by SIs 1982/1727, 1985/67, 1987/2235, 1989/2061 amd 1990/2468.

Regulations as to Pesticide Control

Control of Pesticides Regulations 1986/1510 as amended by SI 1990/2487.

Pesticides (Maximum Residue Levels in Food) Regulations 1988/1378.*

* These regulations will soon be replaced or substantially amended to implement new EC requirements as to pesticide residues.

Regulations as to Food Imitations

Food Imitations (Safety) Regulations 1989/1291.

Orders and Regulations as to Medicinal Claims for Food and Safety of Medicinal Products Affecting Food

Medicines (Exemption from Licences) (Food and Cosmetics) Order 1971/1410 as amended by SI 1973/2079.

Medicines (Retail Sale or Supply of Herbal Remedies) Order 1977/2130.

Medicines (Stilbenes and Thyrostatic Substances Prohibition) Order 1982/518.

Medicines (Stilbenes and Thyrostatic Substances) Regulations 1982/626 as amended by SI 1986/1980.

Regulations and Orders as to the Sale of Food by Quantity

Weights and Measures Act 1963 (Cheese, Fresh Fruits and Vegetables, Meat and Poultry) Order 1984/1315 as amended by SIs 1985/988 and 1985/1980.*

Weights and Measures (Packaged Goods) Regulations 1986/2049 as amended by SI 1987/1538.

Weights and Measures (Quantity Marking and Abbreviations of Units) Regulations 1987/1538 as amended by SI 1988/627.

Weights and Measures (Intoxicating Liquor) Order 1988/2039 as amended by SI 1990/1550.

Weights and Measures (Miscellaneous Foods) Order 1988/2040 as amended by SI 1990/1550.

* Order soon to be replaced.

Orders as to Price Indications

Price Marking (Pre-packed Milk in Vending Machines) Order 1976/796.

Price Marking (Food and Drink on Premises) Order 1979/361.

Price Marking Order 1991/1382.

Special note: For all the above orders and regulations in fully amended and annotated form see *Butterworths Law of Food and Drugs* and *O'Keefe's Law of Weights and Measures* (Butterworths). These volumes are updated three times a year. For up-to-date information as to new or amended legislation see Butterworths monthly *Consumer Law Bulletin*.

Appendix B

Digest of the Principal Instruments of European Community Food Law

The following is a list of the principal EC measures on food law. It does not include numerous Directives and Regulations on marketing, pricing and related matters which are outside the scope of this book. In each case the title, serial number, the Official Journal reference number and, where appropriate, a list of amending instruments, are given together with a brief indication of the scope and content.

Directives do not have the force of law in the United Kingdom until they are implemented by regulations. EEC regulations are immediately binding subject to enforcement arrangements and penalties being introduced by regulations. All implementing regulations etc are listed in Appendix A.

Additives

Council Directive 89/107/EEC (OJ No L 40; 11.02.89) – a framework Directive on the control of additives generally.
Council Directive 88/388/EEC (OJ No L 184; 15.07.88) – a framework Directive on the control of flavourings in food – as completed by Commission Directive 91/71/EEC (OJ No L 42; 15.02.91).

Antioxidants

Council Directive 70/357/EEC (OJ No L 157; 18.07.70) – detailed control and positive list of approved antioxidants-as amended by Council Directives:
73/101 (OJ No L 022; 01.01.73)
74/412 (OJ No L 221; 12.08.74)
78/143 (OJ No L 044; 15.02.78)
81/962 (OJ No L 354; 09.12.81)
85/007 (OJ No L 002; 03.01.85)
87/055 (OJ No L 024; 27.01.87)
Purity Criteria – Council Directive 78/664 (OJ No L 186; 30.06.78).

Colours

Council Directive 2645/62 (OJ No L 115; 11.11.62) – detailed control and positive list of food colours – as amended by Council Directives:

65/469 (OJ No L 178; 26.10.65)
70/358 (OJ No L 157; 18.07.70)
73/101 (OJ No L 002; 01.01.73)
76/399 (OJ No L 108; 26.04.76)
78/144 (OJ No L 044; 15.02.78)
81/712 (OJ No L 257; 10.09.81)
85/007 (OJ No L 002; 03.01.85)
Supplemented by Council Directives:
67/653 (OJ No L 263; 30.10.67)
68/419 (OJ No L 309; 24.12.68)
81/020 (OJ No L 043; 14.02.81).

Emulsifiers and Stabilisers, Thickeners and Gelling Agents

Council Directive 74/329 (OJ No L 189; 12.07.74) – detailed control and positive lists of emulsifiers, stabilisers, thickeners and gelling agents – as amended by Council Directives:
78/612 (OJ No L 197; 22.07.78)
80/597 (OJ No L 155; 23.06.80)
85/006 (OJ No L 002; 03.01.85)
85/007 (OJ No L 002; 03.01.85)
86/102 (OJ No L 088; 03.04.86)
89/393 (OJ No L 186; 30.06.89).
Purity criteria for emulsifiers etc – Council Decision 78/663 (OJ No L 223; 14.08.78) as amended by Council Decision:
82/504 (OJ No L 230; 05.08.82)
and Commission Directive:
90/612 (OJ No L 326; 24.11.90).

Erucic Acid

Council Directive 76/621 (OJ No L 202; 28.07.76) – restriction of the erucic acid content of oil or fat and of food to which oil or fat has been added.

Extraction Solvents

Council Directive 88/344/EEC (OJ No L 157; 24.06.88) – list of extraction solvents which may be used in food and restrictions thereon.

Preservatives

Council Directive 64/54/EEC (OJ No 012; 27.01.64) – detailed control and positive list of preservatives – as amended by Council Directives:
65/569 (OJ No L 222; 28.12.65)
66/722 (OJ No L 233; 20.12.66)
67/427 (OJ No L 148; 11.07.67)
68/420 (OJ No L 309; 24.12.68)
70/359 (OJ No L 359; 18.07.70)
71/160 (OJ No L 087; 17.04.71)
72/002 (OJ No L 002; 01.01.72)
72/444 (OJ No L 298; 31.12.72)
73/101 (OJ No L 002; 01.01.73)
74/062 (OJ No L 038; 11.02.74)
74/394 (OJ No L 208; 30.07.74)
76/462 (OJ No L 126; 14.05.76)

78/145 (OJ No L 044; 15.02.78)
81/214 (OJ No L 101; 11.04.81)
85/007 (OJ No L 002; 03.01.85)
85/172 (OJ No L 065; 06.03.85)
85/585 (OJ No L 372; 31.12.85).

Purity criteria for preservatives: Council Directive 65/66/EEC (OJ No L 022; 09.02.65) as amended and supplemented by Council Directives:
67/428 (OJ No L 148; 11.07.67)
76/463 (OJ No L 126; 14.05.76)
86/604 (OJ No L 352; 13.12.86).

Alcoholic Drinks

Alcoholic Strength

Commission Directive 87/397/EEC (OJ No L 113; 30.04.87) – requirement to declare alcoholic strength on labels.

Marketing Standards for Spirit Drinks

Council Regulation (EEC) No 1576/89 (OJ No L 160; 12.06.89) – definition, description and presentation of spirit drinks.

Detailed Rules for Implementing Regulation 1576/89

Commission Regulation (EEC) No 1014/90 (OJ No L 105; 25.04.90) – as amended by Commission Regulations:
1180/91 (OJ No L 115; 08.05.91)
1781/91 (OJ No L 160; 25.06.91)

and subject to transitional measures in Commission Regulations (EEC) Nos 1759/90 (OJ No L 162; 28.06.90), 3207/90 (OJ No L 307; 07.11.90), and 3750/90 (OJ No L 360; 22.12.90).

Articles Appearing to be Food

Council Directive 87/357/EEC (OJ No L 192; 11.07.87) – prohibition of articles which may be mistaken for food.

Caseins and Caseinates

Council Directive 83/417/EEC (OJ No L 237; 26.08.83) – establishing standards for caseins and caseinates.

Analysis of Caseins and Caseinates

Commssion Directives 85/503 (OJ No L 308; 20.11.85) and 86/424 (OJ No L 243; 28.08.86).

Cocoa and Chocolate Products

Council Directive 73/241/EEC (OJ No L 228; 16.08.73) – establishing standards for cocoa and chocolate products and certain labelling provisions in relation thereto – as amended by Council Directives:

74/411 (OJ No L 221; 12.08.74)
74/644 (OJ No L 349; 28.12.74)
75/155 (OJ No L 064; 11.03.75)
76/628 (OJ No L 223; 16.08.76)
78/609 (OJ No L 197; 22.07.78)
78/842 (OJ No L 291; 17.10.78)
80/608 (OJ No L 170; 03.07.80)
85/007 (OJ No L 002; 03.01.85)
89/344 (OJ No L 142; 03.05.89).

Coffee Extracts and Chicory Extracts

Council Directive 77/436/EEC (OJ No L 172; 12.07.77) – establishing standards for coffee extracts and chicory extracts and certain labelling provisions in relation thereto – as amended by Council Directives:
85/007 (OJ No L 002; 03.01.85)
85/573 (OJ No L 372; 31.12.85).

Analysis of Coffee etc Products

Commission Directive 79/1066/EEC (OJ No L 327; 24.12.79).

Eggs

Marketing Standards

Council Regulation (EEC) No 1907/90 (OJ No L 173; 06.07.90) – establishing marketing standards for eggs.
Commission Regulation (EEC) No 1274/91 (OJ No L 121; 16.05.91) – detailed rules for implementing the above regulation.

Egg Products

Council Directive 89/437/EEC (OJ No L 212; 22.07.89) – hygiene and health problems affecting egg products – as amended by Council Directive:
89/662 (OJ No L 395; 30.12.89).

Enforcement and Administration

Official Control of Foodstuffs

Council Directive 89/397/EEC (OJ No L 186; 30.06.89) – powers of enforcement of Community food law.

Methods of Sampling

Council Directive 85/591/EEC (OJ No L 372; 31.12.85) – methods and sampling and analysis of foodstuffs.

Fish

Organisation of the Market

Council Regulation (EEC) No 3796/81 (OJ No L 379; 31.12.81) – general organisation of the market – this regulation has been amended 12 times but the amendments are not relevant to this work.

Specific Standards

Council Regulation (EEC) No 103/76 (OJ No L 020; 28.01.76) – marketing standards for fresh or chilled fish – as amended by Council Regulations:
0273/81 (OJ No L 030; 02.02.81)
3166/82 (OJ No L 332; 27.1..82)
3250/83 (OJ No L 321; 18.11.83)
3396/85 (OJ No L 322; 03.12.85)
3856/87 (OJ No L 363; 23.12.87)
3940/87 (OJ No L 373; 31.12.87)
0033/89 (OJ No L 005; 07.01.89).

Council Regulation (EEC) No 104/76 (OJ No L 020; 28.01.76) – marketing standards for shrimps – as amended by Council Regulations:
3575/83 (OJ No L 356; 20.12.83)
3118/85 (OJ No L 297; 09.11.85)
3940/87 (OJ No L 373; 31.12.87)
4213/88 (OJ No L 370; 31.12.88)
3162/91 (OJ No L 300; 31.10.91).

Council Regulation (EEC) No 2136/89 (OJ No L 212; 22.07.89) – marketing standards for preserved sardines.

Health Conditions

Council Directive 91/492/EEC (OJ No L 268; 24.09.91) – health conditons for the production and placing on the market of live bivalve molluscs.

Council Directive 91/494/EEC (OJ No L 268; 24.09.91) – health conditions for the production and placing on the market of fishery products.

Foodstuffs for Particular Nutritional Uses

Council Directive 89/398/EEC (OJ No L 186; 30.06.89) – framework Directive empowering the Commission to make vertical Directives in respect of the nutritional benefits of certain foods.

Fresh Fruit and Vegetables

General Marketing Standards

Commission Regulation (EEC) No 1035/72 (OJ No L 118; 20.05.72) – General Marketing Standards. This regulation lays down the general rules for marketing standards and trade in fresh produce. It has been amended 27 times but most amendments are concerned with the general marketing arrangements and are thus outside the scope of this work.

Specific Quality Standards

Specific quality standards are laid down for fruits and vegetables by vertical regulations for:

Apples and pears
Apricots
Artichokes
Asparagus
Beans
Brussels sprouts
Cabbages
Carrots
Caulflowers
Celery
Cherries
Chicory
Cucumbers
Garlic
Grapes
Kiwi fruit
Leeks
Lettuce and endives
Nectarines
Onions
Oranges and other citrus fruit
Peaches
Peas
Plums
Spinach
Strawberries
Sweet peppers
Tomatoes

Fruit Juices and Similar Products

Council Directive 75/726/EEC (OJ No L 311; 01.12.75) – establishing standards for fruit juices and fruit nectars etc and certain labelling provisions in relation thereto – as amended by Council Directives:
79/168 (OJ No L 037; 13.02.79)
81/487 (OJ No L 189; 11.07.81)
89/394 (OJ No L 186; 30.06.89).

Honey

Council Directive 74/409/EEC (OJ No L 221; 12.08.74) – establishing standards for honey and certain labelling provisions in relation thereto.

Jams, Jellies, Marmalades and Chestnut Puree

Council Directive 79/693/EEC (OJ No L 205; 13.08.79) – establishing standards for jams etc and certain labelling provisions in relation thereto – as amended by Council Directives:
80/1276 (OJ No L 375; 31.12.80)
88/593 (OJ No L 318; 25.11.88).

Labelling

Council Directive 79/112/EEC (OJ No L 033; 08.02.79) – laying down detailed Community requirements for the labelling of foodstuffs – as amended by Council Directives:
85/007 (OJ No L 002; 03.01.85)

86/197 (OJ No L 144; 29.05.86)
89/393 (OJ No L 186; 30.06.89)
91/72 (OJ No L 42; 15.02.91).

Temporary Designation of Ingredients

Commission Directive 83/463/EEC (OJ No L 255; 15.09.83).

Nutrition Labelling of Foodstuffs

Council Directive 90/496/EEC (OJ No L 276; 06.10.90).

Batch Marking of Foodstuffs

Council Directive 89/396/EEC (OJ No L 186; 30.06.89) – as amended by Council Directive:
91/238 (OJ No L 107; 27.04.91).

Materials and Article in Contact with Food

General Requirement

Council Directive 89/109/EEC (OJ No L 40; 11.02.89).

Plastics Materials and Articles

Commission Directive 90/128/EEC (OJ No L 349; 13.12.90).

Vinyl Chloride Monomer Materials

Council Directive 78/142/EEC (OJ No L 044; 15.02.78).

Methods and Analysis for Vinyl Chloride Monomers

Commission Directive 80/766/EEC (OJ No L 213; 16.08.80).

Symbol to Accompany Materials etc

Commission Directive 80/590/EEC (OJ No L 151; 19.06.80).

Testing for Migration of Plastics Materials

Council Directive 82/711/EEC (OJ No L 297; 23.10.82).
Council Directive 85/572/EEC (OJ No L 372; 31.12.85).

Regenerated Cellulose Film Materials

Council Directive 83/229/EEC (OJ No L 123; 11.05.83).

Meat

Community measures on the health of live animals are not included in this Appendix.

Fresh Meat

Council Directive 91/497/EEC (OJ No L 268; 24.09.91) – on health problems affecting intra-Community trade in fresh meat to extend it to the production and marketing of fresh meat.

Council Directive 72/461/EEC (OJ No L 302; 31.12.72) – health problems affecting intra-Community trade in fresh meat – as amended by Council Directives:
75/379 (OJ No L 172; 03.07.75)
77/098 (OJ No L 026; 31.01.77)
80/213 (OJ No L 047; 21.02.80)
80/1009 (OJ No L 325; 01.12.80)
81/476 (OJ No L 186; 08.07.81)
82/893 (OJ No L 378; 31.12.82)
83/646 (OJ No L 360; 23.12.83)
84/336 (OJ No L 177; 14.07.84)
84/643 (OJ No L 339; 27.12.84)
85/322 (OJ No L 168; 28.06.85)
87/064 (OJ No L 034; 05.02.87)
87/489 (OJ No L 280; 03.10.87)
89/662 (OJ No L 375; 30.12.89)
91/266 (OJ No L 134; 29.05.91)
and as amended by Council Regulation 3768/85 (EEC) (OJ No L 362; 31.12.85). See also Commission Decision as to bovine spongiform encephalopathy 90/200/EEC (OJ No L 105; 25.04.90).

Council Directive 72/462/EEC (OJ No L 302; 31.12.72) – health and veterinary inspection problems upon importation of bovine animals, swine and fresh meat from third countries – as amended by Council Directives:
75/379 (OJ No L 172; 03.07.75)
77/096 (OJ No L 026; 31.01.77)
77/098 (OJ No L 026; 31.01.77)
78/685 (OJ No L 227; 18.08.78)
81/476 (OJ No L 186; 08.07.81)
83/091 (OJ No L 059; 05.03.83)
86/469 (OJ No L 275; 26.09.86)
87/064 (OJ No L 034; 05.02.87)
88/289 (OJ No L 124; 18.05.88)
88/657 (OJ No L 382; 31.12.88)
89/227 (OJ No L 093; 06.04.89)
90/423 (OJ No L 224; 18.08.90)
90/425 (OJ No L 224; 18.08.90)
90/675 (OJ No L 373; 31.12.90)
91/69 (OJ NO L 046; 19.02.91)
91/266 (OJ No L 134; 29.05.91)
91/496 (OJ No L 268; 24.09.91)
91/497 (OJ No L 268; 24.09.91)
and also as amended by Council Regulation 3768/85 (EEC) (OJ No L 362; 31.12.85).

Council Directive 71/118/EEC (OJ No L 055; 08.03.71) – health problems affecting trade in fresh poultrymeat – as amended by Council Directives:
73/101 (OJ No L 002; 01.01.73)
75/379 (OJ No L 172; 03.07.75)
75/431 (OJ No L 192; 24.07.75)

78/050 (OJ No L 015; 19.01.78)
80/216 (OJ No L 147; 21.02.80)
80/879 (OJ No L 251; 24.09.80)
81/476 (OJ No L 186; 08.07.81)
81/578 (OJ No L 209; 29.07.81)
82/352 (OJ No L 234; 09.08.82)
84/186 (OJ No L 087; 30.03.84)
84/335 (OJ No L 177; 04.07.84)
84/462 (OJ No L 339; 27.12.84)
85/324 (OJ No L 168; 28.06.85)
85/326 (OJ No L 168; 28.06.85)
88/657 (OJ No L 382; 31.12.88)
89/662 (OJ No L 395; 30.12.89)
90/654 (OJ No L 353; 17.12.90)

and as amended by Council Regulations:

3768/85 (OJ No L 362; 31.12.85)
89/662 (OJ No L 357; 31.12.87).

Council Directive 91/494/EEC (OJ No L 268; 24.09.91) – on animal health conditions governing intra-Community trade in and imports from third countries of fresh poultrymeat.

Commission Directive 80/879/EEC (OJ No L 251; 24.09.80) – health marking of large packagings of fresh poultrymeat.

Council Directive 91/495/EEC (OJ No L 268; 24.09.91) – concerning public health and animal health problems affecting the production and placing on the market of rabbit meat and farmed game meat.

Council Directive 91/498/EEC (OJ No L 268; 24.09.91) – on conditions for granting temporary and limited derogations from specific Community health rules on the production and marketing of fresh meat.

Meat Products

Council Directive 91/67/EEC (OJ No L 46; 19.02.91) – on health conditions governing the placing on the market of aquaculture animals and products.

Council Directive 88/657/EEC (OJ No L 382; 31.12.88) – on production of, and trade in, minced meat, meat in pieces of less than 100 grams and meat preparations – as amended by Council Directive:

89/662 (OJ No L 395; 30.12.89).

Note: This Directive has not been implemented by law in the United Kingdom and will be replaced by a new and substantially revised Directive in the near future.

Residues

Council Directive 81/602/EEC (OJ No L 222; 07.08.81) – prohibition of substances having a hormonal and thyrostatic action – supplemented by Council Directive:

85/358 (OJ No L 191; 23.07.85).

Council Directive 86/469/EEC (OJ No L 275; 26.09.86) – examination of animals and fresh meat for the presence of residues.

Milk

Council Regulation (EEC) No 1411/71 (OJ No L 148; 03.07.71) – laying down compositional standards and lawful names for milk as amended by Council Regulations:
74/1556 (OJ No L 167; 22.06.74)
76/566 (OJ No L 067; 15.03.76)
88/222 (OJ No L 028; 01.02.88).

Dehydrated Preserved Milk

Council Directive 76/1128/EEC (OJ No L 024; 30.01.76).

Protection of Dairy Names

Council Regulation (EEC) No 1898/87 (OJ No L 182; 03.07.87) – prevention of the use of milk and dairy names for non-dairy products.

Animal Health Problems Affecting Milk

Council Directive 85/397/EEC (OJ No L 226; 24.08.85) – as amended by Council Directives:
89/662 (OJ No L 395; 30.12.89).
85/3768 (OJ No L 362; 31.12.85)
89/159 (OJ No L 059; 02.03.89)
89/165 (OJ No L 061; 04.03.89)
Commission Regulation (EEC) No 382/89 (OJ No L 44; 16.02.89).
Commission Directive 89/362/EEC (OJ No L 156; 08.06.89).

Natural Mineral Waters

Council Directive 80/777/EEC (OJ No L 229; 30.08.80) – laying down compositional and origin requirements for natural mineral waters and certain labelling requirements in relation theretoas amended by Council Directives:
80/1276 (OJ No L 375; 31.12.80).
85/007 (OJ No L 002; 03.01.85).

Oils and Fats

Marketing Standards

Council Regulation (EEC) No 136/66 (OJ No L 172; 30.09.66) – marketing standards for oils and fats (including olive oil) – as amended by Council Regulations:
2146/68 (OJ No L 314; 31.12.68)
1547/72 (OJ No L 165; 21.07.72)
1707/73 (OJ No L 175; 29.06.73)
2560/77 (OJ No L 303; 28.11.77)
1562/78 (OJ No L 185; 07.07.78)
1585/80 (OJ No L 160; 26.06.80)
1917/80 (OJ No L 186; 19.07.80)
3454/80 (OJ No L 360; 31.12.80)
1413/82 (OJ No L 162; 12.06.82)

1416/82 (OJ No L 162; 12.06.82)
1726/82 (OJ No L 189; 01.07.82)
1097/84 (OJ No L 113; 28.04.84)
1101/84 (OJ No L 113, 28.04.84)
1556/84 (OJ No L 150; 06.06.84)
2260/84 (OJ No L 208; 03.08.84)
231/85 (OJ No L 026; 31.01.85)
3768/85 (OJ No L 362; 31.12.85)
1454/86 (OJ No L 133; 21.05.86)
1915/87 (OJ No L 183; 03.07.87)
3994/87 (OJ No L 377; 31.12.87)
1098/88 (OJ No L 110; 29.04.88)
2210/88 (OJ No L 197; 26.07.88)
1225/89 (OJ No L 128; 11.05.89).
2902/89 (OJ No L 280; 29.09.89)
3499/90 (OJ No L 338; 05.12.90)
1720/91 (OJ No L 162; 26.06.91)

Undesirable Substances

Commission Regulation (EEC) No 983/88 (OJ No L 098; 15.04.88) – prohibition of undesirable substances in oils and fats – as amended by Commission Regulation:
1860/88 (OJ No L 166; 01.07.88).

Pesticide Residues

Fruit and Vegetables

Council Directive 76/895/EEC (OJ No L 340; 09.12.76) – laying down maximum residue levels (MRLs) in fruit and vegatables – as amended by Council Directives:
80/0428 (OJ No L 102; 19.04.80)
80/0036 (OJ No L 046; 19.02.81)
80/1528 (OJ No L 234; 09.08.82)
85/3768 (OJ No L 362; 31.12.85)
88/0298 (OJ No L 126; 20.05.88)
89/186 (OJ No L 66; 10.03.89).

The above regulations are soon to be replaced by new MRLs made under framework Council Directive 90/642/EEC (OJ No L 350; 14.12.90).

Cereals

Council Directive 86/362/EEC (OJ No L 221; 07.08.86) – maximum MRLs for cereals – as amended by Council Directive:
88/0298 (OJ No L 126; 20.05.88).
90/654 (OJ No L 353; 17.12.90).

Foodstuffs of Animal Origin

Council Directive 86/363/EEC (OJ No L 221; 07.08.86).

Poultry and Poultrymeat

Marketing Standards

Council Regulation (EEC) No 1906/90 (OJ No L 173; 06.07.90) – marketing standards for poultry carcases etc.
Commission Regulation 1538/91/EEC (OJ No L 143; 07.06.91) – detailed rules as to marketing standards.

Water Content

Council Regulation (EEC) No 2967/76 (OJ No L 339; 08.12.76) – laying down standards for the water content of frozen poultrymeat – as amended by Council Regulations:
1691/77 (OJ No L 188; 28.07.77)
0641/79 (OJ No L 082; 31.03.79)
2632/80 (OJ No L 270; 15.10.80)
2785/80 (OJ No L 288; 31.10.80)
2835/80 (OJ No L 292; 01.11.80)
3204/83 (OJ No L 315; 15.11.83).

Detailed Rules for Implementing above

Commission Regulation 2785/80 (OJ No L 288; 31.10.80) – as amended by Commission Regulations:
3134/81 (OJ No L 312; 31.10.81)
3759/85 (OJ No L 356; 31.12.85).

Product Liability

Council Directive 85/374/EEC (OJ No L 210; 07.08.85) – establishing a regime of strict product liability.

Radioactive Contamination

Council Regulation (Euratom) No 3954/87 (OJ No L 371; 30.12.87) as amended by Council Regulations (Euratom):
2218/89 (OJ No L 211; 22.07.89)
2219/89 (OJ No L 211; 22.07.89).

Sugars

Council Directive 73/437/EEC (OJ No L 356; 27.12.73) – laying down certain standards for sugars and sugar products.

Analysis of Sugars

Commission Directive 79/796/EEC (OJ No L 239; 22.09.79).

Quick-Frozen Foodstuffs

Council Directive 89/108/EEC (OJ No L 40; 11.02.89) – temperature, storage and quality requirements as to foodstuffs marked 'quick-frozen'.

Veterinary Medicinal Product Residues

Council Regulation (EEC) No 2377/90 (OJ No L 224; 18.08.90) framework Directive for controlling medicinal product residues in food.

Wine

Analysis

Commission Regulation (EEC) No 2676/90 (OJ No L 272; 03.10.90) – Community methods of analysis of wine products.

Common Organisation of the Market in Wine

Council Regulation (EEC) No 822/87 (OJ No L 084; 27.03.87) – as amended by Council Regulations:
1390/87 (OJ No L 133; 22.05.87)
1972/87 (OJ No L 184; 03.07.87)
3146/87 (OJ No L 300; 23.10.87)
3992/87 (OJ No L 377; 31.12.87)
1441/88 (OJ No L 132; 28.05.88)
2253/88 (OJ No L 198; 26.07.88)
2964/88 (OJ No L 269; 29.09.88)
4252/88 (OJ No L 373; 31.12.88)
1236/89 (OJ No L 128; 11.05.89
0388/90 (OJ No L 042; 06.02.90)
1325/90 (OJ No L 132; 23.05.90)
3577/90 (OJ No L 353; 17.12.90)
1594/91 (OJ No L 148; 13.06.91)
1734/91 (OJ No L 163; 26.06.91).

Description and Presentation

Council Regulation (EEC) No 2392/89 (OJ No L 232; 09.08.89) – as amended by Council Regulation:
3886/89 (OJ No L 378; 27.12.89)
3827/90 (OJ No L 366; 29.12.90)
2356/91 (OJ No L 216; 03.08.91).

Commission Regulation (EEC) No 3201/90 (OJ No L 309; 08.11.90) – detailed rules for the description of wine and grape musts.

Council Regulation (EEC) No 3309/85 (OJ No L 320; 29.11.85) – description of sparkling wines and aerated sparkling wines – as amended by Council Regulations:
3805/85 (OJ No L 367; 31.12.85)
1626/86 (OJ No L 144; 29.05.86)
0538/87 (OJ No L 055; 25.02.87)
2045/89 (OJ No L 202; 14.07.89)
2357/91 (OJ No L 214; 03.08.91).

Council Regulation (EEC) No 2707/86 (OJ No L 246; 30.08.86) – detailed rules for sparkling and aerated sparkling wines) – as amended by Commission Regulations:
3778/86 (OJ No L 319; 05.11.86)
2249/87 (OJ No L 207; 29.07.87)

0575/88 (OJ No L 56; 02.03.88)
2657/88 (OJ No L 237; 27.08.88)
0596/89 (OJ No L 065; 09.03.89)
2776/90 (OJ No L 267; 29.09.90)
3826/90 (OJ No L 366; 29.12.90).

Aromatised Wines etc

Council Regulation (EEC) No 1601/91 (OJ No L 149; 14.06.91) – general rules for aromatised wines, aromatised wine-based drinks and aromatised wine product cocktails.
Commission Regulation (EEC) No 3664/91 (OJ No L 348; 17.12.91) – transitional measures re Council Regulation (EEC) No 1601/91 above.

Enriching, Acidifying and Deacidifying

Commission Regulation (EEC) No 2094/86 (OJ No L 180; 04.07.86) – rules for use of tartaric acid for deacidification of certain wine products as amended by Commission Regulation:
2736/86 (OJ No L 252; 04.09.86).
Commission Regulation (EEC) No 2240/89 (OJ No L 215; 26.07.89) – monitoring the use of enriching and deacidifying of wine.

Sweetening

Commission Regulation (EEC) No. 1618/70 (OJ No L 175; 08.08.70).

Control of Quality Wines

Council Regulation (EEC) No 2046/89 (OJ No L 202; 14.07.89) – general rules on controls.
Commission Regulation (EEC) No 1698/70 (OJ No L 190; 26.08.70) – certain derogations on production – as amended by Commission Regulations:
0101/73 (OJ No L 002; 01.01.73)
0807/73 (OJ No L 078; 27.03.73)
0986/89 (OJ No L 106; 18.04.89).
Commission Regulation (EEC) No 2247/73 (OJ No L 230; 18.08.73) – quality wines produced in specified regions – as amended by Commission Regulations:
0418/86 (OJ No L 048; 26.02.86)
0986/89 (OJ No L 106; 18.04.89).
Commission Regulation (EEC) No 3590/83 (OJ No L 363; 24.12.83) – methods of analysis for neutral alcohol as amended by Commission Regulations:
1614/86 (OJ No L 142; 28.05.86)
2039/88 (OJ No L 179; 09.07.88).
Council Regulation (EEC) No 823/87 (OJ No L 084; 27.03.87) – quality wines produced in specified regions – as amended by Council Regulation:
2043/89 (OJ No L 202; 14.09.89).
Council Regulation (EEC) No 4252/88 (OJ No L 373; 31.12.88) – preparation and marking of liqueur wines as amended by Council Regulations:
1328/90 (OJ No L 132; 23.05.90)
1735/91 (OJ No L 163; 26.06.91).

Commission Regulation (EEC) No 2903/79 (OJ No L 326; 22.12.79) – downgrading of quality wines produced in specified regions-as amended by Commission Regulation:
0418/86 (OJ No L 048; 26.02.86).
Commission Regulation (EEC) No 3590/85 (OJ No L 343; 20.12.85) – certificate of analysis required for imports.
Commission Regulation (EEC) No 1888/86 (OJ No L 163; 19.06.86) – maximum sulphur dioxide content of certain sparkling wines.

List of Approved Wines

Commission Regulation (EEC) No 2805/73 (OJ No L 230; 18.08.73) – list of quality white wines – as amended by Commission Regulations:
3548/73 (OJ No L 361; 29.12.73)
2160/75 (OJ No L 220; 20.08.75)
0966/77 (OJ No L 115; 06.05.77).
Commission Regulation (EEC) No 2082/74 (OJ No L 217; 08.08.74) – list of quality liqueur wines produced in specified regions.

Coupage

Commission Regulation (EEC) No 479/86 (OJ No L 54; 01.03.86) – coupage of red Spanish wines with red wines from other member states.
Commission Regulations (EEC) No 1781/86 (OJ No L 155; 10.06.86) and 2202/89 (OJ No L 209; 21.07.89) – Detailed rules for coupage with Spanish red wines.
Commission Regulation (EEC) No 1781/86 (OJ No L 155; 10.06.86) and Council Regulation (EEC) No 0479/86 (OJ No L 054; 01.03.86) – Coupage with Spanish red wines of other red wines.
Commission Regulation (EEC) No 643/77 (OJ No L 081; 30.03.77) – wines from third countries – as amended by Commission Regulations:
3203/80 (OJ No L 333; 11.12.80)
0418/86 (OJ No L 048; 26.02.86).

Oenological Practices

Commission Regulation (EEC) No 1972/78 (OJ No L 226; 17.08.78).

Prescribed Forms

Commission Regulation (EEC) No 1153/75 (OJ No L 113; 01.05.75) – wine producers and traders – as amended by Council Regulations:
2617/77 (OJ No L 304; 29.11.77)
3203/80 (OJ No L 333; 11.12.80)
0418/86 (OJ No L 048; 26.02.86)
0986/89 (OJ No L 106; 18.04.88).
Commission Regulation (EEC) No 986/89 (OJ No L 106; 18.04.89) – Accompanying documents for carriage of wines and records to be kept – as amended by Council Regulation:
2246/90 (OJ No L 203; 01.09.90).

Controlled Maximum Alcoholic Strength

Commission Regulation (EEC) No 2319/74 (OJ No L 248; 11.09.74) – Wine growing areas which may produce table wines have a prescribed maximum alcoholic strength.

Council Regulation (EEC) No 351/79 (OJ No L 054; 05.03.79) – addition of alcohol to wines – as amended by Council Regulations:
3658/81 (OJ No L 366; 22.12.81)
0255/87 (OJ No L 026; 29.01.87)
4090/87 (OJ No L 382; 31.12.87)
3904/88 (OJ No L 347; 16.12.88)
1372/90 (OJ No L 133; 24.05.90).

Indication of Alcoholic Strength

Council Regulation (EEC) No 1627/86 (OJ No L 144; 29.05.86) – general rules.

Council Regulation (EEC) No 1069/87 (OJ No L 104; 16.04.87) – indications on special wines.

Council Regulation (EEC) No 3677/89 (OJ No L 360; 09.12.89) – total alcoholic strength and total acidity of imported wines.

Sparkling Wines

Commission Regulation (EEC) No 2152/75 (OJ No L 219; 19.08.75) – detailed rules – as amended by Commission Regulation:
0986/89 (OJ No L 106; 18.04.89).

Council Regulation (EEC) No 358/79 (OJ No L 054; 05.03.79) – Community produced wines – as amended by Council Regulations:
2383/79 (OJ No L 274; 31.10.79)
3456/80 (OJ No L 360; 31.12.80)
3686/84 (OJ No L 341; 29.12.84)
3310/85 (OJ No L 320; 29.11.85)
3805/86 (OJ No L 367; 31.12.85)
2044/89 (OJ No L 202; 14.07.89).

Imports

Commission Regulation (EEC) No 1393/76 (OJ No L 157; 18.06.86) – detailed rules on importation of wine products from third countries – as amended by Commission Regulations:
0688/78 (OJ No L 093; 07.04.78)
1666/78 (OJ No L 192; 15.07.78)
1243/79 (OJ No L 158; 26.06.79)
3104/80 (OJ No L 324; 29.11.80)
3671/89 (OJ No L 358; 08.12.89).

Council Regulation (EEC) No 2390/89 (OJ No L 232; 09.12.89) – general rules for the importation of wines, grape juice and grape must – as amended by Council Regulations:
3887/87 (OJ No L 378; 27.12.89)
2179/90 (OJ No L 198; 28.07.90)
3331/91 (OJ No L 316; 16.11.91).

Note: There are numerous regulations on the classification of vine varieties and registers for such purposes which are outside the scope of this work.

Appendix C

Permitted Additives in Food

The following additives are authorised for use in food (see Chapter 7). The list is up-to-date to 1 January 1992.

The following tables should not be used as a substitute for the regulations.

Antioxidants

The Antioxidants in Food Regulations 1978/105, as amended (see Appendix A). Council Directive 70/537/EEC, as amended (see Appendix B).

Permitted Antioxidants

Serial Number	*Name and synonyms*
E300	L-Ascorbic acid
E301	Sodium L-ascorbate
E302	Calcium L-ascorbate
E304	6-o-Palmitoyl-L-ascorbic acid (ascorbyl palmitate)
E306	Extracts of natural origin rich in tocopherols
E307	Synthetic *alpha*-tocopherol
E308	Synthetic *gamma*-tocopherol
E309	Synthetic *delta*-tocopherol
E310	Propyl gallate (propyl 3,4,5-trihydroxybenzoate)
E311	Octyl gallate (octyl 3,4,5-trihydroxybenzoate)
E312	Dodecyl gallate (dodecyl 3,4,5-trihydroxybezoate)
E320	Butylated hydroxyanisole (BHA)
E321	Butylated hrdroxytoluene (BHT (2,6-di-*tert*-butyl-p-cresol)
–	Ethoxyquin

Antioxidants Permitted Only in Certain Foods

Food	*Permitted antioxidant*	*Maximum mg/kg*
Any of the following dairy products for use in the preparation of food: Butter, butter fat, dried cream, dried cheese, dried whey, dried whey derivatives.	(1) (2) (3)	100(4) 200(4) 300(4,5)

Food	*Permitted antioxidant*	*Maximum mg/kg*
Anhydrous edible oils and fats and vitamin oils and concentrates other than preparations of Vitamin A or Vitamin A esters containing more than 30,000 μg retinol equivalents per gram.	(1) (2) (3)	100 200 300(5)
Any permitted emulsifier or stabiliser containing combined fatty acids whether or not those fatty acids have been polymerised.	(1) (2) (3)	100 200 300(5)
Essential oils and isolates from the concentrates of essential oils.	(1) (2) (3)	1000 1000 1000
Potato powder, potato flakes and potato granules.	(2)	25
Walnuts (shelled)	E320 E321 (2)	70 70 140(6)
Apples and pears.	Ethoxyquin	3
Preparations of Vitamin A and Vitamin A esters containing more than 30,000 μg retinol equivalents per gram.	(2)	33.3(7)
Preparations of Vitamin A and Vitamin A esters for use as ingredients in the preparation of food described directly or indirectly as intended for babies or young children.	(2)	0.35(7)
Chewing gum base.	(2)	1000
Chewing gum manufactured from chewing gum base containing BHT.	(2)	200

(1) E310, E311, E312 or any mixture of them
(2) E320 or E321 or any mixture of them
(3) Any mixture of (1) or (2)
(4) Calculated on the milk fat content
(5) Of which not more than 100 shall be from (1) and not more than 200 shall be from (2)
(6) Of which not more than 70 shall be BHA and not more than 70 shall be BHT
(7) mg/kg of preparation for every 1000 μg retinol equivalents per gram.

Notes:
(1) The following are not permitted in foods intended for babies or young children (except as specifically permitted above) – E320, E321, E310, E311, E312, ethoxyquin.
(2) Diphenylamine may be used on or in apples and pears up to a maximum of 10 mg/kg where it is present solely as a scald inhibitor.

(3) Any food (except specified dairy products) containing as an ingredient one of the above foods may contain antioxidant in proportion to its use.

Colours

The Colouring Matter in Food Regulations 1973/1340, as amended (see Appendix A). Council Directive 2645/62/EEC, as amended (see Appendix B).

Permitted Colours:

Serial number	*Name of colour*	*Colour index number**
E100	Curcumin	75300
E101	Riboflavin or Lactoflavin	–
–	Riboflavin-5'-phosphate	–
	Riboflavin-5'-sodium phosphate	–
E102	Tartrazine	19140
E104	Quinolene Yellow	47005
E110	Sunset yellow FCF or Orange Yellow S	15985
E120	Cochineal or Carminic acid	75470
E122	Carmoisine or Azorubine	14720
E123	Amaranth	16185
E124	Ponceau 4R or Cochineal Red A	16255
E127	Erythrosine BS	45430
–	Red 2G	18050
E131	Patent Blue V	42051
E132	Indigo Carmine or Indigotine	73015
--	Brilliant Blue FCF	42090
E140	Chlorophyll	75810
E141	Copper complexes of chlorophyll and chlorophyllins	75810
E142	Green S or Acid Brilliant Green BS Lissamine Green	44090
–	Brown FK	–
–	Chocolate Brown HT	20285
E150	Caramel	–
E151	Black PN or Brilliant Black BN	28440
E153	Carbon Black or Vegetable Carbon	–
E160	Carotenoids:	
E160(a)	alpha-carotine, beta-carotine, gamma carotine	75130
E160(b)	annatto, bixin, norbixin	75120
E160(c)	capsanthin or capsorubin	–
E160(d)	lycopene	75125
E160(e)	beta-apo-8'-carotenal (C30)	40820
E160(f)	ethyl ester of beta-apo-8'-carotenoic acid (C30)	40825
E161(a)	Flavoxanthin	–

Serial number	*Name of colour*	*Colour index number**
E161(b)	Lutein	–
E161(c)	Cryptoxanthin	–
E161(d)	Rubixanthin	75135
E161(e)	Violaxanthin	–
E161(f)	Rhodoxanthin	–
E161(g)	Canthaxanthin	40850
E162	Beetroot Red or Betanin	–
E163	Anthocyanins	–
–	The following natural substances having a secondary colouring effect:	
	(a) paprika	–
	(b) turmeric	75300
	(c) saffron	75100
	(d) sandalwood	–
	(e) the pure colouring principle of any of the colouring matters listed under (a) to (d) above	
E171	Titanium dioxide	77891
E172	Iron oxides and hydroxides	77489 77491 77492 77499
E173	Aluminium**	77000
E174	Silver**	–
E175	Gold**	77480
E180	Pigment Rubine or Lithol Rubine BK***	15850
–	Methyl violet****	42535
–	The synthetic equivalent identical with the pure colouring principle of any natural colouring matter described above	

Notes:

* Colour Index reference numbers are allocated in the Colour Index (CI) of the Society of Dyers and Colourists (3rd Edn).

** For use in dragees for external colouring only and in sugar coated flour confectionery for decoration only.

*** For use on the rind of hard cheese only.

**** For marking raw or unprocessed meat only.

Emulsifiers and Stabilisers

Emulsifiers and Stabilisers in Food Regulations 1989/876, as amended (see Appendix A); Council Directive 74/329/EEC, as amended (see Appendix B).

Permitted Emulsifiers and Stabilisers:

Serial number	*Emulsifier or stabiliser*
E322	Lecithins
E400	Alginic acid

Serial number	*Emulsifier or Stabiliser*
E401	Sodium alginate
E402	Potassium alginate
E403	Ammonium alginate
E404	Calcium alginate
E405	Propane-1,2-diol alginate
E406	Agar
E407	Carrageenan
E410	Locust bean gum
E412	Guar gum
E413	Tragacanth
E414	Acacia
E415	Xanthan gum
416	Karaya gum
432	Polyoxyethylene (20) sorbitan monolaurate
433	Polyoxyethylene (20) sorbitan mono-oleate
434	Polyoxyethylene (20) sorbitan monopalmitate
435	Polyoxyethylene (20) sorbitan monostearate
436	Polyoxyethylene (20) sorbitan tristearate
E440	(i) Pectin
E440	(ii) Amidated pectin
442	Ammonium phosphatides
E460	(i) Microcrystalline cellulose
E461	Methylcellulose
E463	Hydroxypropylcellulose
E464	Hydroxypropylmethylcellulose
E465	Ethylethylcellulose
E466	Carboxymethylcellulose
E470	Sodium, potassium and calcium salts of fatty acids
E471	Mono- and di-glycerides of fatty acids
E472(a)	Acetic acid esters of mono- and di-glycerides of fatty acids
E472(b)	Lactic acid esters of mono- and di–glycerides of fatty acids
E472(c)	Citric acid esters of mono- and di-glycerides of fatty acids
E472(d)	Tartaric acid esters of mono- and di-glycerides of fatty acids
E472(e)	Mono- and diacetyltartaric acid esters of mono- and di-glycerides of fatty acids
E472(f)	Mixed acetic and tartaric acid esters of mono- and di-glycerides of fatty acids
E473	Sucrose esters of fatty acids
E474	Sucroglycerides
E475	Polyglycerol esters of fatty acids
476	Polyglycerol esters of polycondensed fatty acids of caster oil

Serial number	*Emulsifier or Stabiliser*
E477	Propane-1,2-diol esters of fatty acids
E481	Sodium stearoyl-2-lactylate
E482	Calcium stearoyl-2-lactylate
E483	Stearyl tartrate
491	Sorbitan monostearate
492	Sorbitan tristearate
493	Sorbitan monolaurate
494	Sorbitan mono-oleate
495	Sorbitan monopalmitate
–	Extract of Quillaia
–	Polyglycerol esters of dimerised fatty acids of soya bean oil
–	Oxidatively polymerised soya bean oil
–	Pectin extract

Emulsifiers and Stabilisers Permitted Only in Certain Food:

Permitted emulsifier or stabiliser	*Specified food*	*Maximum mg/kg*
(a) Extract of Quillaia	Soft drinks	200(1)
(b) Sodium, potassium and calcium salts of fatty acids:		
(i) all such salts	Dutch type rusks	15,000(2)
	The permitted emulsifiers or permitted stabilisers – mono- and di-glycerides of fatty acids; lactic acid esters of mono and di-glycerides of fatty acids; sucrose esters of fatty acids; sucroglycerides; polyglycerol esters of fatty acids; propane 1,2-diol esters of fatty acids	60,000(3)
(ii) soldium octanoate	Malter barley	(4)
(c) Oxidatively polymerised soya bean oil Polyglycerol esters of dimerised fatty acids of soya bean oil	Tin greasing emulsions	(4)

(1) of the dry matter content of the extract
(2) calculated as sodium oleate, $C_{18}H_{33}O_2Na$ and on the weight of the flour
(3) calculated as sodium oleate, $C_{18}H_{33}O_2Na$
(4) in accordance with good manufacturing practice

Permitted Emulsifiers and Stabilisers in Food Subject to Other Regulations

The Bread and Flour Regulations 1984 impose restrictions on the use of emulsifiers and stabilisers in bread. There are similar restrictions in the Cheese Regulations

1970 in respect of soft cheese, whey cheese, processed cheese and cheese spread and in the Cocoa and Chocolate Products Regulations 1976 in respect of cocoa products and chocolate products.

Mineral Hydrocarbons

The Mineral Hydrocarbons in Food Regulations 1966/1073, as amended (see Appendix A) prohibit the use of mineral hydrocarbons in the composition of preparation of food with certain exemptions for dried fruit, citrus fruit, sugar confectionery, chewing compounds, and the rind of pressed cheese and the shells of eggs. Ministers have announced their intention to revoke these regulations and to prohibit the use of mineral hydrocarbons in or on all foods. Consequent upon that announcement the use of mineral hydrocarbons, which may be carcinogenic, has been voluntarily discontinued by the food industry.

Miscellaneous Additives

Miscellaneous Additives in Food Regulations 1980, as amended (see Appendix A); Council Directives Nos 2645/62/EEC, 64/54/EEC, 65/66/EEC, 70/357/EEC, 78/612/EEC, 78/663/EEC and 78/664/EEC, as amended (see Appendix B).

Permitted Miscellaneous Additives:

Serial number	*Name of miscellaneous additive*
E260	Acetic acid
–	Sodium acetate, anhydrous
–	Sodium acetate
E262	Sodium hydrogen diacetate
E261	Potassium acetate
E263	Calcium acetate
–	Adipic acid
–	2-Aminoethanol
E927	Azodicarbonamide
–	Beeswax, white
–	Beeswax, yellow
–	Benzoyl peroxide
–	Calcium phytate
E290	Carbon dioxide
–	Ammonium carbonate
–	Ammonium hydrogen carbonate
–	Sodium carbonate
–	Sodium hydrogen carbonate
–	Sodium sesquicarbonate
–	Magnesium carbonate, heavy
–	Magnesium carbonate, light
–	Potassium carbonate
–	Potassium hydrogen carbonate
E170	Calcium carbonate
–	Carnauba wax

Serial number	*Name of miscellaneous additive*
E460(ii)	*Alpha*-cellulose
E925	Chlorine
E926	Chlorine dioxide
E330	Citric acid
–	*tri*Ammonium citrate
E331	Sodium dihydrogen citrate
E331	*di*Sodium citrate
E331	*tri*Sodium citrate
E332	Potassium dihydrogen citrate
E332	*tri*Potassium citrate
E333	*mono*Calcium citrate
E333	*di*Calcium citrate
E333	*tri*Calcium citrate
–	Ammonium ferric citrate
–	Ammonium ferric citrate, green
E920	L-Cysteine hydrochloride
–	Dichlorodiflouromethane
–	Dimethypolysiloxane
–	*di*Sodium dihydrogen ethylenediamine – NNN'N' tetra-acetate
–	Calcium disodium ethylenediamine – NNN'N' tetra-acetate
–	Sodium ferrocyanide
–	Potassium ferrocyanide
–	Fumaric acid
–	D-Glucono-1,5-lactone
–	Sodium gluconate
–	Potassium gluconate
–	Calcium gluconate
–	Sodium hydrogen L-glutamate
–	Glycine
–	1,4-Heptonolactone
–	Sodium heptonate
–	Calcium heptonate
–	Hydrocholoric acid
–	Ammonium chloride
–	Potassium chloride
–	Calcium chloride, anhydrous
–	Calcium chloride
–	Hydrogen
–	Ammonium hydroxide
–	Sodium hydroxide
–	Magnesium hydroxide
–	Magensium oxide, heavy
–	Magnesium oxide, light
–	Potassium hydroxide
–	Calcium hydroxide
–	Calcium oxide
E270	Lactic acid
E325	Sodium lactate
E326	Potassium lactate
E327	Calcium lactate

Serial number	*Name of miscellaneous additive*
–	DL-Malic acid
–	L-Malic acid
–	Sodium hydrogen malate
–	Sodium malate
–	Potassium malate
–	Calcium hydrogen malate
–	Calcium malate
–	Metatartaric acid
–	Nicotinic acid
–	Nitrogen
–	Nitrous oxide
–	Octadecylammonium acetate
–	Oxygen
–	Oxystearin
E338	Orthophosphoric acid
–	Ammonium dihydrogen orthophosphate
–	*di*Ammonium hydrogen orthophosphate
E339	Sodium dihydrogen orthosphosphate
E339	*di*Sodium hydrogen orthophosphate
E339	*tri*Sodium orthophosphate
E340	Potassium dihydrogen orthophosphate
E340	*di*Potassium hydrogen orthophosphate
E340	*tri*Potassium orthophosphate
E341	Calcium tetrahydrogen diorthophosphate
E341	Calcium hydrogen orthophosphate
E341	*tri*Calcium diorthophosphate
–	Sodium aluminium phosphate, acidic
–	Sodium aluminium phosphate, basic
E450(a)	*di*Sodium dihydrogen diphosphate
E450(a)	*tri*Sodium diphosphate
E450(a)	*tetra*Sodium diphosphate
E450(a)	*tetra*Potassium diphosphate
–	*di*Calcium diphosphate
E450(b)	*penta*Sodium triphosphate
E450(b)	*penta*Potassium triphosphate
E450(c)	Sodium polyphosphates
E450(c)	Potassium polyphosphates
–	Ammonium and calcium polyphosphates
–	Edible bone phosphate
–	Guanosine 5'-(disodium phosphate)
–	Inosine 5'- (disodium phosphate)
–	Polydextrose
–	Sodium 5'-ribonucleotide
–	Shellac
–	Silicon dioxide
–	Bentonite
–	Kaolin, heavy
–	Kaolin, light
–	Aluminium sodium silicate
–	Aluminium calcium silicate
–	Calcium silicate
–	Magnesium silicate, synthetic

Serial number	*Name of miscellaneous additive*
–	Magnesium trisilicate
–	Talc
–	Spermaceti
–	Sperm oil
–	Magnesium stearate
–	Calcium stearate
–	Butyl stearate
–	Succinic acid
–	Sulphuric acid
–	Ammonium sulphate
–	Sodium sulphate
–	Magnesium sulphate
–	Potassium sulphate
–	Aluminium potassium sulphate
–	Calcium sulphate
–	Tannic acid
E334	L-(+)-Tartaric acid
–	DL-Tartaric acid
E335	*mono*Sodium L-(+)-tartrate
–	*mono*Sodium DL-tartrate
E335	*di*Sodium L-(+)-tartrate
–	*di*Sodium DL-tartrate
E336	*mono*Potassium L-(+)–tartrate
–	*mono*Potassium DL-tartrate
E336	*di*Potassium L-(+)-tartrate
–	*di*Potassium DL-tartrate
E337	Potassium sodium L-(+)-tartrate
–	Potassium sodium DL-tartrate

Miscellaneous Additives permitted Only in Certain Foods:

Specified food	*Permitted miscellaneous additive*	*Except where otherwise stated max mg/kg*
Ammonium chloride	Octadecylammonium acetate	500
Brandy	*di*Sodium dihydrogen ethylenediamine NNN'N'-tetra acetate	25 mg/l
Bread	Azodicarbonamide Benzol peroxide Chlorine dioxide L-Cysteine hydrochloride	As prescribed in the Bread & Flour Regulations 1984
Canned fish and canned fish	Calcium disodium ethylenediamine-NNN'N'-tetra acetate	In accordance with good manufacturing practice
Chocolate confectionery	Carnauba wax	200
Chocolate products	Carnauba wax	200

Specified food	*Permitted miscellaneous additive*	*Except where otherwise stated max mg/kg*
Flour	Azodicarbonamide Benzoyl peroxide Chlorine Chlorine dioxide L-Cysteine hydrochloride	As prescribed by the Bread & Flour Regulations 1984
Frozen food	Dichlorodifluoro-methane	100 (when thawed at and to 20°C)
Glace cherries	Calcium disodium ethylenediamine-NNN'N'-tetra-acetate Aluminium potassium sulphate	In accordance with good manufacturing practice 10,000 (dry matter basis)
Peeled fruit	2-Aminoethanol	100
Peeled vegetables	2-Aminoethanol	100
Sugar confectionery	Carnauba wax	200
Wine	Metatartaric acid	100 mg/l

Food for Babies and Young Children May Not Contain:

2-aminoethanol, *alpha*-cellulose, sodium hydrogen L-glutamate, guanosine 5'-(disodium phosphate), inosine 5'-(disodium phosphate), polydextrose or sodium 5'-ribonucleotide.

Preservatives

The Preservatives in Food Regulations 1989, as amended (see Appendix A); Council Directive 64/54/EEC, as amended (see Appendix B).

Permitted Preservatives:

Permitted preservative	*Serial number*	*Alternative form**	*Serial number*
Sorbic acid	E200	Sodium sorbate Potassium sorbate Calcium sorbate	E201 E202 E203
Benzoic acid	E210	Sodium benzoate Potassium benzoate Calcium benzoate	E211 E212 E213
Ethyl 4-hydroxybenzoate	E214	Ethyl 4-hydroxybenzoate sodium salt	E215
Propyl 4-hydroxybenzoate	E216	Propyl 4-hydroxybezoate sodium salt	E217
Methyl 4-hydroxybenzoate	E218	Methyl 4-hydroxybenzoate sodium salt	E219

Permitted preservative	*Serial number*	*Alternative form**	*Serial number*
Sulphur dioxide	E220	Sodium sulphite	E221
		Sodium hydrogen sulphite	E222
		Sodium metabisulphite	E223
		Potassium metabisulphite	E224
		Calcium sulphite	E226
		Calcium hydrogen sulphite	E227
Potassium bisulphate	E228		
Biphenyl	E233		
2-Hydroxybiphenyl	E231	Sodium biphenyl-2-oxide	E232
2-(Thiazol-4-yl) benzimdazole	E233		
Hexamine	E239		
Sodium nitrite	E250	Potassium nitrite	E249
Sodium nitrate	E251	Potassium nitrate	E252
Propionic acid	E280	Sodium propionate	E281
		Calcium propionate	E282
		Potassium propionate	E283
Nisin	234		

Articles of Food which may Contain permitted Preservative and the Nature and Proportion of Permitted Preservative in each case:

Specified food	*Permitted preservative*	*Max mg/kg unless otherwise stated*
Beer	E220 and either	70
	E210 or	70
	E215 or	70
	E217	70
Beetroot, cooked and prepacked	E210 or	250
	E219 or	250
	E215	250
Bread	E280	As prescribed by the Bread & Flour Regulations 1984
Cauliflower, canned	E220	100
Cheese	E200	1,000
Cheese other than Cheddar, Cheshire, Grana-padano or Provolone type cheeses or soft cheese	E251 and E250	50, of which not more than 5 may be E250 expressed in both cases as E250
Provolone cheese	E239	25 (expressed as formaldehyde)

Specified food	*Permitted preservative*	*Max mg/kg unless otherwise stated*
Chicory and coffee essence	E210 or	450
	E218 or	450
	E215 or	450
	E216	450
Christmas pudding	E280	1,000
Cider	E220 or	200
	E200	200
Coconut, desiccated	E220	50
Colouring matter, except E150 caramel, if in the form of a solution of a permitted colouring matter	E210 or	2,000
	E218 or	2,000
	E214 or	2,000
	E216 or	2,000
	E200	1,000
Colouring matter E150 caramel	E220	1,000
Crabmeat, canned	E220	30
Desserts, fruit-based milk and cream	E220 or	100
	E200	300
Dessert sauces, fruit based with a total soluble solids content of less than 75%	E220 or	100
	E210 or	250
	E218 or	250
	E214 or	250
	E216 or	250
	E200	250
		1,000
The permitted miscellaneous additive, dimethylpolysiloxane	E220 or	1,000
	E210 or	2,000
	E218 or	2,000
	E214 or	2,000
	E216 or	2,000
	E200	1,000
Enzymes:		
Papain, solid	E220	30,000
Papain, aqueous solutions	E220 or	5,000
	E220	1,000
Aqueous solutions of enzyme preparations not otherwise specified including immobilised enzyme preparations in aqueous media	E220 or	500
	E210 or	3,000
	E218 or	3,000
	E214 or	3,000
	E216 or	3,000
	E200	3,000
Fat spreads consisting of an emulsion principally of water in		

Specified food	*Permitted preservative*	*Max mg/kg unless otherwise stated*
oil with a fat content not exceeding 70%	E200	2,000
Figs, dried	E220 or	2,000
	E200	500
Fillings and toppings for flour confectionery which consist principally of a sweetened oil and water emulsion with a min sugar solids content of 50%	E200	1,000
Finings when sold by retail:		
Wine finings	E220	12,500
Beer finings	E220	50,000
Flavouring syrups	E220 or	350
	E210 or	800
	E218 or	800
	E214 or	800
	E216	800
Flavourings	E220 or	350
	E210 or	800
	E218 or	800
	E214 or	800
	E216	800
Flour confectionery	E280 or	1,000
	E200	1,000
Foam headings, liquid	E220 or	5,000
	E210 or	10,000
	E218 or	10,000
	E214 or	10,000
	E216	10,000
Freeze drinks	E220 or	70
	E210 or	160
	E218 or	160
	E214 or	160
	E216 or	160
	E200	300
Fruit-based pie fillings	E220 or	350
	E210 or	800
	E218 or	800
	E214 or	800
	E216 or	800
	E200	450

Specified food	*Permitted preservative*	*Max mg/kg unless otherwise stated*
Fruit, dried, other than prunes or figs	E220	2,000
Fruit, fresh:		
Bananas	E233	3
Citrus fruit	E230 or E231 or E233	70 12 10
Grapes	E220	15
Fruit, fruit pulp, or fruit puree (including tomatoes, tomato pulp, tomato paste and puree) which in each case is not fresh or canned	E220 or E210 or E218 or E214 or E216	350 800 800 800 800
Fruit juices:		
Any fruit juice or concentrated fruit juice mentioned in reg 11 (2) of the Fruit Juices and Fruit Nectars Regs 1977	E220	As prescribed in the regulations
Any other fruit juice or concentrated fruit juice	E220 or E210 or E218 or E214 or E216	350 800 800 800 800
Fruit or plants (including flowers and seeds), crystallised, glace or drained (syruped), or candied peel or cut and drained (syruped) peel	E220 or E210 or E218 or E214 or E216 or E200	100 1,000 1,000 1,000 1,000 1,000
Fruit pieces in stabilised syrup for use as ingredients of ice cream or other edible ices	E200	1,000
Fruit spread	E220 and E200	100 1,000
Garlic, powdered	E220	2,000
Gelatin	E220	1,000
Gelatin capsules	E200	3,000
Ginger, dry root	E220	150

Specified food	*Permitted preservative*	*Max mg/kg unless otherwise stated*
Glucose drinks containing not less than 235 g of glucose syrup per litre of drink	E220 or E210 or E218 or E214 or E216	350 800 800 800 800
Grape juice products (unfermented intended for sacramental use)	E220 and E210 or E218 or E214 or E216	70 2,000 2,000 2,000 2,000
Grape juice, concentrated intended for home wine making and labelled as such	E220	2,000
Hamburgers or similar products	E220	450
Herring, marinated whose pH does not exceed 4.5	E210 or E218 or E214 or E216	1,000 1,000 1,000 1,000
Herring, marinated whose pH exceeds 4.5	E239 and E210 or E218 or E214 or E216	50 1,000 1,000 1,000 1,000
Hops, dried, sold by retail	E220	2,000
Horseradish, fresh grated and horseradish sauce	E220 or E210 or E218 or E214 or E216	200 250 250 250 250
Jam and similar products:		
Extra jam, jam, extra jelly, jelly, marmalade, chestnut puree, UK standard jelly, X curd, Y flavour curd, mincemeat;	E220	As prescribed in the Jam & Similar Products Regs 1981
Reduced sugar jam, reduced sugar jelly, and reduced sugar marmalade	E220 and E210 or E218 or E214 or E216 or E200	" " "

Specified food	*Permitted preservative*	*Max mg/kg unless otherwise stated*
Makerel, marinated – whose pH does not exceed 4.5	E210 or E218 or E214 or E216	1,000 1,000 1,000 1,000
Makerel, marinated – whose pH exceeds 4.5	E239 and E210 or E218 or E214 or E216	1,000 1,000 1,000 1,000 1,000
Mallow, chocolate covered	E200	1,000 calc on wt of mallow and chocolate
Meat, cured meat (including cured meat products) packed in a sterile pack	E251 and E250	150 of which not more than 50 may be E250
Acidified and/or fermented cured meat products (including salami and similar products) not packed in a sterile pack	E251 and E250	400 of which not more than 50 may be E250
Uncooked bacon and ham; cooked bacon and ham that is not and has not been packed in any hermetically sealed container	E251 and E250	500 of which not more than 200 may be E250
Any cured meat or cured product not mentioned above	E251 and E250	250 of which not more than 150 may be E250
Mushrooms, frozen	E220	50
Nut pastes, sweetened	E200	1,000
Olives, pickled	E220 or E210 or E218 or E214 or E216 or E200	100 250 250 250 250 500
Peas, garden, canned containing no added colouring matter	E220	100
Pectin, liquid	E220	250
Perry	E220 or E200	200 200

Specified food	*Permitted preservative*	*Max mg/kg unless otherwise stated*
Pickles, other than pickled olives	E220 or E210 or E218 or E214 or E216 or E200	100 250 250 250 250 1,000
Potatoes, raw peeled	E220	50
Prawns, shrimps and scampi, other than prawns and shrimps in brine	E220	200 in the edible part
Prawns and shrimps in brine	E220 and either E200 and either E210 or E214 or E216 or E218	200 in the edible part 2,000 2,000 300 300 300
Preparations of saccharin, sodium saccharin or calcium saccharin and water only	E210 and either E218 or E214 or E216	750 250 250 250
Prunes	E200 or E220	1,000 2,000
Salad cream (incl mayonnaise) and salad dressing	E220 or E210 or E218 or E214 or E216 or E200	100 250 250 250 250 1,000
Sambal oelek	E210 and E200	850 1,000
Sauces, other than horseradish sauce	E220 or E210 or E218 or E214 or E216 or E200	100 250 250 250 250 1,000
Sausage or sausage meat	E220	450
Snack meals, concentrated (moisture 15 to 60%)	E220 and E218	1,500 175

Specified food	*Permitted preservative*	*Max mg/kg unless otherwise stated*
Soft drinks for consumption after dilution not otherwise specified in this table	E220 or E210 or E218 or E214 or E216 or E200	350 800 800 800 800 1,500
Soft drinks for consumption without dilution not otherwise specified in this table	E220 or E210 or E218 or E214 or E216 or E200	70 160 160 160 160 300
Soup concentrates (moisture 25 to 60%)	E200 and E218	1,500 175
Starches, including modified starches	E220	100
Sugars:		
Specified sugar products	E220	As per the Specified Sugar Products Regs 1976
Hydrolised starches (other than specified sugar products)	E220	400
Other sugars except lactose	E220	70
Tea extract liquid	E210 or E218 or E214 or E216	450 450 450 450
Vegetables, dehydrated:		
Brussel sprouts	E220	2,500
Cabbage	E220	2,500
Potato	E220	550
Others	E220	2,000
Vinegar:		
Cider or wine vinegar	E220	200
Other	E220	70
Wine (including alcoholic cordials) other than Community-controlled wine	E220 and E200	450 mg/l 200 mg/l

Specified food	*Permitted preservative*	*Max mg/kg unless otherwise stated*
Yogurt, fruit	E220 or	60
	E210 or	120
	E218 or	120
	E214 or	120
	E216 or	120
	E200	300

Notes:
(1) Reg 4 of the Preservatives in Food Regulations 1989 provides for various maximum quantities in which alternative preservatives or admixtures of preservatives are permitted in the above table.
(2) Food for babies or young children may not contain sodium nitrate or sodium nitrite.

Permitted Preservatives which are Controlled by Other Regulations:

For the use of permitted preservatives in bread, see the Bread and Flour Regulations 1984; for such use in fruit juices and fruit nectars see the Fruit Juices and Fruit Nectars Regulations 1977; for such use on preserves see the Jam and Similar Products Regulations 1981; and for use in sugars, see the Specified Sugar Products Regulations 1976. For wines, the use of preservatives is subject to Community regulations for which see Appendix B.

Solvents

The Solvents in Food Regulations 1967/1582 as amended (see Appendix A).

Permitted Solvents:

Serial number	*Name of permitted solvent*
–	Ethyl alcohol, ethanol
–	Ethyl acetate
–	Diethyl ether, solvent ether
E422	Glycerol
–	Glycerol mono-acetate (monoacetin)
–	Glycerol di-acetate (diacetin)
–	Glycerol tri-acetate (triacetin)
–	Iso-propyl alcohol
–	Propylene glycol

Sweeteners

The Sweeteners in Food Regulations 1983/1211, as amended (see Appendix A).

Permitted Sweeteners:

Serial number	*Name*
–	Acesulfame potassium
–	Aspartame
–	Hydrogenated glucose syrup
–	Isomalt
–	Lactitol

Serial number	*Name*
E421	Mannitol
–	Saccharin
–	Sodium saccharin
–	Calcium saccharin
E420	Sorbitol
E420	Sorbitol syrup
–	Thaumatin
–	Xylitol

Permitted Additives Given Temporary Serial Numbers

Food Labelling Regulations 1984, Schedule 5.

Name of additive	*Serial number*
Riboflavin-5'-phosphate	101a
Yellow 2G	107
Red 2G	128
Brilliant Blue FCF	133
Brown FK	154
Chocolate Brown HT	155
Nisin	234
Sodium acetate, anhydrous Sodium acetate	262
DL-Malic acid L-Malic acid	296
Fumaric acid	297
Sodium Malate Sodium hydrogen malate	350
Potassium malate	351
Calcium malate Calcium hydrogen malate	352
Metatartaric acid	353
Adipic acid	355
Succinic acid	363
1,4-Heptonolactone	370
Nicotinic acid	375
*tri*Ammonium citrate	380
Ammonium ferric citrate Ammonium ferric citrate green	381
Calcium disodium ethylenediamine-NNN'N' tetra-acetate (Synonym: Calcium disodium EDTA)	385
Karaya Gum	416
Polyoxyethylene (8) stearate	430
Polyoxyethylene (40) stearate	431
Polyoxyethylene (20) sorbitan monolaurate	432
Polyoxyethylene (20) sorbitan mono-oleate	433
Polyoxyethylene (20) sorbitan monopalmate	434
Polyoxyethylene (20) sorbitan monostearate	435
Polyoxyethylene (20) sorbitan tristearate	436
Ammonium phosphatides	442

Name of additive	*Serial number*
Polyglycerol esters of polycondensed fatty acids of castor oil (Synonym: Polyglycerol polyricinoleate)	476
Lactylated fatty acid esters of glycerol and propane-1, 2-diol	478
Sorbitan monostearate	491
Sorbitan tristearate	492
Sorbitan monolaurate	493
Sorbitan mono-oleate	494
Sorbitan monopalmitate	495
Sodium carbonate Sodium hydrogen carbonate Sodium sesquicarbonate	500
Potassium carbonate Potassium hydrogen carbonate	501
Ammonium carbonate Ammonium hydrogen carbonate	503
Magnesium carbonate, heavy Magnesium carbonate, light	504
Hydrochloric acid	507
Potassium chloride	508
Calcium chloride, anhydrous Calcium chloride	509
Ammonium chloride	510
Sulphuric acid	513
Sodium sulphate	514
Potassium sulphate	515
Calcium sulphate	516
Magnesium sulphate	518
Sodium hydroxide	524
Potassium hydroxide	525
Calcium hydroxide	526
Ammonium hydroxide	527
Magnesium hydroxide	528
Calcium oxide	529
Magnesium oxide, heavy Magnesium oxide, light	530
Sodium ferrocyanide	535
Potassium ferrocyanide	536
*di*Calcium diphosphate	540
Sodium aluminium phosphate, acidic Sodium aluminium phosphate, basic	541
Edible bone phosphate	542
Calcium polyphosphates	544
Ammonium polyphosphates	545
Silicon dioxide	551
Calcium silicate	552
Magnesium silicate, synthetic Magnesium trisilicate	553a
Talc	553b
Aluminium sodium silicate	554
Aluminium calcium silicate	556

Name of additive	*Serial number*
Bentonite	558
Kaolin, heavy Kaolin, light	559
Stearic acid	570
Magnesium stearate	572
D-Glucono-1,5-lactone (Synonym: Glucono *delta*-lactone)	575
Sodium gluconate	576
Potassium gluconate	577
Calcium gluconate	578
L-glutamic acid	620
Sodium hydrogen L-glutamate (Synonym: *mono* Sodium glutamate)	621
Potassium hydrogen L-glutamate (Synonym: *mono*Potassium glutamate)	622
Calcium dihydrogen di-L-glutamate (Synonym: Calcium glutamate)	623
Guanosine 5'-(disodium phosphate) (Synonym: Sodium guanylate)	627
Inosine 5'-(disodium phosphate) (Synonym: Sodium inosinate)	631
Sodium 5'-ribonucleotide	635
Maltol	636
Ethyl maltol	637
Dimethylpolysiloxane	900
Beeswax, white Beewswax, yellow	901
Carnauba wax	903
Shellac	904
Mineral hydrocarbons	905
Refined microcrystalline wax	907
L-cysteine hydrochloride L-cysteine hydrochloride monohydrate	920
Potassium bromate	924
Chlorine	925
Chlorine dioxide	926
Azodicarbonamide (Synonym: Azoformamide)	927

Appendix D

Some Useful Addresses

Advertising Standards Authority,
Brook House, 2–16 Torrington Place, London, WC1E 7HN
Tel: 071 580 5555

British Meat Manufacturers Association,
19 Cornwall terrace, London, NW1 4QP
Tel: 071 935 7980

British Standards Institution,
2 Park Street, London, W1A 2BS
Tel: 071 629 9000

Butterworth Law Publishers Ltd,
Borough Green, Sevenoaks, Kent, TN15 8PH
Tel: 0732 884567

Campden Food Preservation Research Association,
Chipping Campden, Gloucestershire, GL55 6LD
Tel: 0386 840319

Commission of the European Communities,
8 Story's Gate, London, SW1P 3AT
Tel: 071 222 8122

Department of Agriculture & Fisheries for Scotland,
Pentland House, Robb's Loan, Edinburgh, EH14 1TW
Tel: 031 556 8400

Department of Health,
Eileen House, 80–94 Newington Causeway, London SE1 6EF
Tel: 071 972 2000

Department of Health
Medicines Division, Market Towers, 1 Nine Elms Lane, London, SW8 5NQ
Tel: 071 720 2188

Department of Trade & Industry,
Consumer Affairs Division, 10–18 Victoria Street, London, SW1H 0NN
Tel: 071 215 7877

Food & Drink Federation,
6 Catherine Street, London WC2B 5JJ

Flour Milling & Baking Research Association,
Chorleywood, Rickmansworth, Hertfordshire, WD3 5SH
Tel: 0923 284111

Government Chemist,
Queens Road, Teddington, Middlesex, TW11 0LY
Tel: 081 943 7000

Health and Safety Executive,
1 Chepstow Place, Westbourne Grove, London W2 4TF
Tel: 071 229 3456

Home Office,
F2 Division, Queen Anne's Gate, London, SW1H 9AT
Tel: 071 273 3000

Independent Television Companies Association,
Code of Advertising Standards and Practice; Advertising and Sponsorship Division,
70 Brompton Road, London, SW3 1EY
Tel: 071 584 7011

Institution of Environmental Health Officers,
Chadwick House, Rushworth Street, London SE1 0QT
Tel: 071 928 6006

Institute of Food Science and Technology,
5 Cambridge Court, 210 Shepherds Bush Road, London, W6 7NL
Tel: 071 603 6316

Institute of Trading Standards Administration,
4/5 Hadleigh Business Centre, 35 London Road, Hadleigh, Essex, SS7 2BT
Tel: 0702 559922

Local Authorities Co-ordinating Body on Trading Standards,
PO Box 6, Token House, 1A Robert Street, Croydon, CR9 1LG
Tel: 081 688 1996

Ministry of Agriculture, Fisheries and Food,
Ergon House, c/o Nobel House, 17 Smith Square, London SW1P 3HX
Tel: 071 238 3000

Ministry of Agriculture, Fisheries and Food, Meat Hygiene Division, Tolworth Tower, Surbiton, Surrey KT6 7DX
Tel: 081 330 4411

National Consumer Council,
20 Grosvenor Gardens, London, SW1W
Tel: 071 730 3469

National Farmers' Union,
Agriculture House, Knightsbridge, London, SW1X 7NJ
Tel: 071 235 5077

National Weights and Measures Laboratory,
Department of Trade & Industry, Stanton Avenue, Teddington, Middlesex, TW11 0JZ
Tel: 081 943 7272

Office of Fair Trading,
Field House, 15–25 Bream's Buildings, London, EC4A 1PR
Tel: 071 269 8800

Retail Consortium,
Bedford House, 69/79 Fulham High Street, London SW6 3JW
Tel: 071 371 5185

Welsh Office,
Cathays Park, Cardiff, CF1 3NQ
Tel: 0222 823515

Law Laboratories Ltd,
121 Shady Lane, Great Barr, Birmingham, B44 9ET
Tel: 021 366 6000

Leatherhead Food Research Association,
Randalls Road, Leatherhead, Surrey, KT22 7RY
Tel: 0372 376761

Index

Location references are to paragraph numbers, figures in bold type indicating chapters. References to text in the appendices are prefixed by the abbreviation 'App'. Abbreviations used in the index are as follows:

COT	Committee on the Toxicity of Chemicals in Food, Consumer Products and the Environment
CPA	Consumer Protection Act
EC	European Community
ECJ	European Court of Justice
FAC	Food Advisory Committee
FEPA	Food and Environment Protection Act
FSA	Food Safety Act
PARNUTS	foods for particular nutritional uses
PCEA	Police and Criminal Evidence Act
QUID	quantitative indication of ingredients
RDA	recommended daily amount
SEA	Single European Act
UK	United Kingdom
WMA	Weights and Measures Act